T0331562

A Fortunate Universe

Life in a Finely Tuned Cosmos

Over the last 40 years, scientists have uncovered evidence that if the Universe had been forged with even slightly different properties, life as we know it – and life as we can imagine it – would be impossible.

Join us on a journey through how we understand the Universe, from its most basic particles and forces, to planets, stars and galaxies, and back through cosmic history to the birth of the cosmos. Conflicting notions about our place in the Universe are defined, defended and critiqued from scientific, philosophical and religious viewpoints. The authors' engaging and witty style addresses what fine-tuning might mean for the future of physics and the search for the ultimate laws of nature.

Tackling difficult questions and providing thought-provoking answers, this volume challenges us to consider our place in the cosmos, regardless of our initial convictions.

GERAINT F. LEWIS is a professor of astrophysics at the Sydney Institute for Astronomy, part of the University of Sydney. With an undergraduate education at the University of London, and a Ph.D. in astrophysics from the world-renowned Institute of Astronomy at the University of Cambridge, Professor Lewis is an internationally recognized astrophysicist, having published more than 200 papers in a diverse range of fields, including gravitational lensing, galactic cannibalism, cosmology and large-scale structure. As well as being an accomplished lecturer, he regularly engages in public outreach through public speaking, articles in the popular press, and through social media, on twitter as @Cosmic_Horizons and on his blog at cosmic-horizons.blogspot.com.

LUKE A. BARNES is a postdoctoral researcher at the Sydney Institute for Astronomy. His university medal from the University of Sydney helped Dr Barnes earn a scholarship to complete a Ph.D. at the University of Cambridge. He has published papers in the field of galaxy formation and on the fine-tuning of the Universe for life. He has been invited to speak at the 2011 and 2015 St Thomas Summer Seminars in Philosophy of Religion and Philosophical Theology, the University of California Summer School for the Philosophy of Cosmology, and numerous public lectures. He blogs at letterstona ture.wordpress.com and tweets @lukebarnesastro.

'My colleagues, Geraint and Luke, take you on a tour of the Cosmos in all of its glory, and all of its mystery. You will see that humanity appears to be part of a remarkable set of circumstances involving a special time around a special planet, which orbits a special star, all within a specially constructed Universe. It is these sets of conditions that have allowed humans to ponder our place in space and time. I have no idea why we are here, but I do know the Universe is beautiful. *A Fortunate Universe* captures the mysterious beauty of the Cosmos in a way that all can share.'

Brian Schmidt, Australian National University, Canberra; Nobel Laureate in Physics (2011)

'Geraint Lewis and Luke Barnes provide a breath-taking tour of contemporary physics from the subatomic to the cosmological scale. Everywhere they find the Universe to be fine-tuned for complex structure. If the quark masses, or the basic forces, or the cosmological constant had been much different, the Universe would have been a sterile wasteland. It seems that the only reactions are either to embrace a multiverse or a designer. The authors have constructed a powerful case for the specialness of our Universe.'

Tim Maudlin, New York University

'The Universe could have been of such a nature that no life at all could exist. The anthropic question asks why the constants of nature that enter various physical laws are such as to permit life to come into being. This engaging book is a well-written and detailed explanation of all the many ways these physical constants affect the possibility of life, considering atomic, nuclear and particle physics, astrophysics and cosmology. It then discusses in an open minded way the variety of explanations one might give for this strange fine-tuning, possible solutions ranging from pure chance, existence of multiverses, or theistic explanations. The book is the most comprehensive current discussion of this intriguing range of issues. Highly recommended.'

George Ellis, University of Cape Town, South Africa

'Lewis and Barnes' book is the most up-to-date, accurate, and comprehensive explication of the evidence that the Universe is fine-tuned for life. It is also among the two most philosophically sophisticated treatments, all the while being accessible to a non-academic audience. I strongly recommend this book.'

Robin Collins, Messiah College, Pennsylvania

'... charming, intelligent and exceedingly well-written ... a gentle stroll through the details of the Standard Model of particle physics, as well as the Standard Model of cosmology, but [the authors] lead us with such a light hand, a streak of humour and a lack of pedantry that the information is easily absorbed ... Lewis and Barnes show us how small changes lead to a variety of disasters. ('Ruining a universe is easy' Mr. Barnes quips) ... Is [our universe] a happy coincidence, as the authors ask each other in an amusing mock debate modeled on one Galileo wrote 400 years earlier, or is there some deeper reason? Where does science go from here? Does what has been popularly called a theory of everything exist? Is there a multiverse? Must we be satisfied with an anthropic principle? The authors discuss these questions and more in a final dialogue.'

Gino Segrè, The Wall Street Journal

'*A Fortunate Universe: Life in a Finely Tuned Cosmos* by Geraint Lewis and Luke Barnes, is a nice up to date book for the general (educated) public on modern physics and cosmology. If covers modern cosmology and some of the Big Questions of our times, in particular the issue of anthropomorphism how 'fine-tuned' our Universe is.'

Steinn Sigurðsson, ScienceBlogs
(www.scienceblogs.com)

'... what is truly unique about this book is that it presents the data at a popular level so that the material is accessible to anyone interested in this topic ... As I read the book, I was awestruck by the finely tuned constants and conditions that had to be just right to get a universe that would permit life ... This evidence should move each one of us to ask, what is the best explanation of this incredible fine-tuning?'

Tim Barnett, Stand to Reason (www.str.org)

'*A Fortunate Universe* is basically a book of physics, written by two scientists who are fascinated by the question 'Why are we here?' The language is straightforward, the style is easy, often witty, with short digestible paragraphs, and yet the subject-matter is inevitably dense and demanding ... It is pleasing to come across the line 'we do not know' so regularly in this book about the fundamentals of science, which echoes the book of Job ... When science reaches its limits, we have to consider a different kind of explanation for why the laws of nature are as they are, and why they are so finely tuned for the emergence of intelligent life. ... [The authors] wonder if classical arguments for the existence of God have anything to say about the fine-tuning of the universe, speculating whether God is a necessary being and whether our sense of truth and morality hint at God's inevitable existence.

Adam Ford, Church Times

'The title claims that the Universe is finely tuned for the existence of life. The authors provide evidence for this, investigate various possible explanations, and rebut the most common criticisms ... the book provides an opportunity to learn more at an accessible level ... The case is well made that the Universe is finely tuned for life; the interesting question is why. It could be coincidence ... Or could the Universe be no other way? ... Was it designed? Did it evolve? Or are there many universes in a Multiverse, and we shouldn't be surprised that we live in one which allows life? ... The arguments are clear; references are provided for those wishing to delve deeper; essentially all points of view are presented ... This is an important topic and the book is a good summary of the field. I enjoyed reading it and recommend it to those interested in the Big Question.'

Phillip Helbig, The Observatory

'It is the vivid, direct tone and writing style of a friendly physics lecture that perhaps most sets this text apart among popular-level science books about 'big questions' ... [The book] provides a big picture of the physics of fine-tuning, mostly accessible in lay terms, and gives aspiring philosophers of physics a taste of the tone and intellectual style one can find at cosmology conferences. Beyond that, it does so by showing the readers that a response from philosophers might be welcome. Because the authors make clear how their thinking is informed by works in metaphysics, philosophy of physics, epistemology, and the philosophy of religion, they tell the readers how they think philosophy does or could contribute, and where they think they do not know enough to see how it might.'

Yann Benétreau-Dupin,
Notre Dame Philosophical Reviews

'This book is for anyone who has ever wondered: 'Why is it so?' With colourful analogies and admirably accurate simplifications, Geraint and Luke have succeeded in making much of modern physics and cosmology comprehensible ... They address the biggest questions of science. What is dark energy? What is dark matter? Why is there something rather than nothing? Why is there more matter than antimatter? Where did the laws physics come from? Do we live in a multiverse? Do we live in a simulation? How different could the universe have been? If God is omnipotent, why does evil exist? ... Not even the popular scientist and writer Paul Davies tries to address so many important big questions in one book ... I enjoyed the book a lot, but I disagreed with the main thesis. No matter what your religious beliefs are, this book will make you think.

Charley Lineweaver, The Conversation
(www.theconversation.com)

A Fortunate Universe

Life in a Finely Tuned Cosmos

GERAINT F. LEWIS
University of Sydney

LUKE A. BARNES
University of Sydney

Foreword by
BRIAN SCHMIDT
Australian National University

University Printing House, Cambridge CB2 8BS, United Kingdom

Cambridge University Press is part of the University of Cambridge.

It furthers the University's mission by disseminating knowledge in the pursuit of education, learning, and research at the highest international levels of excellence.

www.cambridge.org
Information on this title: www.cambridge.org/9781108747400
DOI: 10.1017/9781316661413

First published 2016
First paperback edition 2020

A catalogue record for this publication is available from the British Library.

Library of Congress Cataloguing in Publication Data
Lewis, Geraint F. | Barnes, Luke A., 1983– | Schmidt, Brian, 1967–
A fortunate universe : life in a finely-tuned cosmos / Geraint F. Lewis, University of Sydney, Luke A. Barnes, University of Sydney, Brian Schmidt, Australian National University.
Cambridge : Cambridge University Press, 2017. | Includes bibliographical references and index.
LCCN 2016029411 | ISBN 9781107156616
LCSH: Pattern formation (Physical sciences) | Pattern formation (Biology) | Life – Origin. | Universe. | Cosmology – Philosophy.
LCC Q172.5.C45 L4845 2017 | DDC 576.8/3–dc23
LC record available at https://lccn.loc.gov/2016029411

ISBN 978-1-107-15661-6 Hardback
ISBN 978-1-108-74740-0 Paperback

Geraint

To slightly misquote The Mamas and the Papas, this is dedicated to the ones we love.

Luke

To slightly misquote R.E.M., this is dedicated to the ones we love.

Contents

Foreword

Like a Bach fugue, the Universe has a beautiful elegance about it, governed by laws whose mathematical precision is meted out to the metronome of time. These equations of physics are finely balanced, with the constants of nature that underpin the equations tuned to values that allow our remarkable Universe to exist in a form where we, humanity, can study it. A slight change to these constants, and poof, in a puff of gedanken experimentation, we have a cosmos where atoms cease to be, or where planets are unable to form. We seem to truly be fortunate to be part of Our Universe.

A seemingly perfectly rational argument to come to terms with this streak of good luck is that, since we exist, we must therefore live in a Universe where we can exist. But this idea has at its heart the notion that ours is selected from a multitude of universes – and there is no evidence for, or against, such a construct of nature.

Our Universe is the only one we have, and this presents a remarkable problem for those of us who study it. Why is it the way it is? Science is founded on using ideas, often called theories, to make predictions. But what happens when, as with our Universe, there is only one thing to observe? Is a theory able to make a prediction when it is either right or wrong on one count?

My colleagues, Geraint and Luke, in *A Fortunate Universe*, take you on a tour of the cosmos in all of its glory, and all of its mystery. Along the way you will learn about the fundamental equations of quantum mechanics that govern our existence, about the concepts behind energy and entropy (and don't be fooled by their description of Canberra, which has far more free energy in it than any Sydney-sider will ever realize), and of course about gravity, which is the primary governor of the Universe on planetary and larger scales.

On your journey with Geraint and Luke, you will see that humanity appears to be part of a remarkable set of circumstances involving a special time around a special planet, which orbits a special star, all within a specially constructed Universe. It is this set of conditions that has allowed humans to ponder our place in space and time. I have no idea why we are here, but I do know the Universe is beautiful. *A Fortunate Universe* captures the mysterious beauty of the cosmos in a way that all can share.

Brian Schmidt
Australian National University
Canberra

Preface

To a human, living on Earth feels just right. Of course, many of the human race face challenges, such as poverty and sickness, on a daily basis, but it can feel like our planet was made for us.

We find ourselves in a neatly placed orbit around a stable, middle-aged star, with the strength of our bones nicely matched by the Earth's gravitational pull, allowing us to ramble freely over the planet. There is oxygen to breathe, and we can power ourselves through the digestion of many tasty plants and animals that inhabit the surface. We would last but a few seconds if we were dropped onto our neighbouring planets. We would be crushed and roasted on the surface of Venus, or left gasping and freezing in Mars's tenuous atmosphere. For humans, the Earth is a special place, a relative cosmic paradise where the conditions are just right for life, including our human life.

But over the past few centuries, we've come to discover how we came to be so nicely suited to conditions on Earth. Our physical properties, our bone structure, our organs, our senses, result from life continually changing and evolving over the last 3.5 billion years, adapting to the conditions that surround us.

The realization that the Earth is not unique changes our view of our place in the Universe. Driven by the continual advancement of science, we have found that humans are part of the web of life, that the Earth is just one of a myriad of planets, and the Sun is but a boringly typical star. Our place in the Universe is just like many, many others, and in no way unique.

Peering more deeply into these same scientific advances, examining the basic make-up of the Universe, reveals that we are not as mediocre as it seems. The fundamental particles from which everything is constructed, and the fundamental forces that dictate interactions, appear to be fine-tuned for life. Minor tinkering with either would leave the Universe dead and sterile.

With every step forward in science, these fine-tuning issues have become more significant. We find ourselves questioning the nature of many of the things we take for granted, from the fabric of space and time, to the mathematical underpinnings of the Universe. At every level, we find that our Universe's ability to create and sustain life forms is rare and remarkable.

The discussion of this cosmological fine-tuning for life has found a very broad audience, from philosophers and physicists in the halls of academia, to religious believers who see the mysterious hand of the divine. It has captured the attention of the popular media, and generated random frothings in various recesses of the internet. All too frequently, the science, and what it is actually telling us about the fine-tuning of the Universe for life, is lost in the noise.

The goal of this book is to present the scientific viewpoint of the fine-tuning of the laws of science, and delve into its implications for the inner workings of the Universe. We will call upon the latest academic and philosophical musings to clarify what fine-tuning actually means and to set the scene for what we can conclude from our existence as life forms.

This book has been a long time in gestation, with the original idea coming from many rambling conversations between the authors and others, the kind of discussions and arguments that lie at the heart of science. Sitting around the table, we wondered about the expansion of the Universe, the nature of electrons, and how many different kinds of universes there could be. We scratched our heads over the make-up of dark matter and dark energy, and wondered deeply about how things could have been different. This quickly leads to the realization that life would be very difficult, if not impossible, in the vast sea of possible universes.

Our hope is that this book crystallizes these discussions, reflecting the rollercoaster of the scientific journey. We hope it gets you thinking about the question that drove us, the question that has dogged humans from the earliest times, the question that we hope we are on the road to answering: why are we here?

Acknowledgements

'Why don't you?' With these words, this book was born. They were uttered by science communicator extraordinaire, Dr Karl Kruszelnicki, when Geraint stated that he had always wanted to write a book. Since this initial conversation, Dr Karl has been a continual source of information, inspiration and enthusiasm.

Writing this book presented a challenge. But the act of getting this book into print presented a mystery, especially to two cosmologists with no understanding of the book industry. But a coffee at the lovely Michaelhouse Cafe with Vince Higgs from Cambridge University Press set us firmly on the road to publishing. His support and professionalism to bring us to this point have been exemplary, and he deserves our greatest thanks.

Many colleagues, both near and far, contributed to the development of this book. Thanks to Mike Irwin and Rodrigo Ibata for astronomical images, and Pascal Elahi for cosmological simulations. There are many more with whom we have chatted, argued and harangued over the question of fine-tuning, far too many to name here. We hope we have convinced you that this apparently trivial problem is not as trivial as it might seem.

For the brave souls who volunteered to read early drafts of this book, we thank you. Thank you Nick Bate, Jon Sharp and the anonymous CUP reviewers. We would also like to thank Robin Collins, Trent Dougherty, Allen Hainline, Osame Kinouchi, Tom Murcko, Matt Payne, Josh Rasmussen, Brad Rettler, Mike Rota, Daniel Rubio and Stuart Starr.

Friends are a vital part of life, and Geraint thanks Matt and Jon for the sporadic meetings over the last thirty years, meetings that have led to much laughter and adventure. Rodrigo is thanked for his friendship and intellectual jousts, from physics to economics, history to biology, and many, many hours debating Gott's Doomsday hypothesis.

Through our families we have received immeasurable love and support. Words can be inadequate, but Bryn and Dylan, you have been the most important and wonderful things in Geraint's world from your first seconds on this planet. Except for Zdenka, you are simply more wonderful and more important. To my parents and brother, I hope this book explains what I actually do for a living!

Luke would also like to thank Geraint for inviting him to be a co-author, and for his dependable and unique brand of grumpy enthusiasm. A special thank you to all the audiences who have interacted with his talks about fine-tuning, and especially the philosophers at the 2011 and 2015 St Thomas Summer Seminars in Philosophy of Religion, run by Mike Rota and Dean Zimmerman. Luke is supported by a grant from the John Templeton Foundation. The opinions expressed in this publication are those of the author and do not necessarily reflect the views of the John Templeton Foundation.

Luke thanks Bernadette; to have a wife who is willing to support book writing, travelling, Saturday afternoon cricket, and bass ukulele playing is 'worth far more than rubies'. You are amazing. To my kids, for being the cutest 5- and 2-year-olds on the planet, and insisting on a cuddle, kiss and high-five every morning before I leave for work. To my parents and siblings, for their constant love and support; 'photons in a box' are explained in detail in Chapter 6.

If there are any words we can give to budding authors, it's seize the day, open that crisp new document, send those emails, and remember that fortune truly does favour the brave.

1   A Conversation on Fine-Tuning

You don't have to be a scientist to appreciate the beauty of the night sky, but there is much more to the Universe[1] than its good looks. For scientists, the goal is to unveil the inner workings of nature, the rules and properties that dictate how the bits and pieces of the cosmos move and interact.

After several centuries of scientific progress, centuries that have revealed so much about our cosmos's fundamental forces and building blocks, science is facing a seemingly simple question whose answer could completely change what we think about the physical world. And that question is 'Why is the Universe just right for the formation of complex, intelligent beings?' This might seem to be a strange question: of course our Universe (or at least, this part of it) is hospitable to human life ... we're here, aren't we? But, could it have been different? And how different could it have been? Could the Universe have been completely sterile and devoid of life?

You may be asking yourself 'how could the Universe have been different?' and the answer is the fundamental laws of its matter and energy could have been different. Our best, deepest theories of physics, which describe how the Universe behaves, have a few loose ends. For all the predictive power of these laws, there are basic quantities that theorists cannot calculate; we have to cheat by getting the answer from experiments. These loose ends cry out for a deeper understanding.

Like writers of alternative history novels, we can ask hypothetical questions about the Universe. Specifically, how different would

---

[1] Throughout this book, our Universe, the one we actually inhabit, will appear capitalized, while hypothetical universes will appear in lower-case.

FIGURE 1 A cake recipe illustrates fine-tuning. You can slightly vary the amounts of the ingredients and still make a tasty cake. But deviate too far, add too many extra ingredients, or leave too many ingredients out, and an inedible mess results.

the Universe have been if it were born with a different set of fundamental properties?

These hypothetical universes may not be significantly different from our own, and so we could guess that they too would be hospitable to human life. Or they could be radically different, but still allow an alternative form of life.

But what if almost all of the possible universes are sterile, with conditions too simple or extreme for life of any conceivable type to arise? Then we are faced with a conundrum. Why, in the almost infinite sea of possibilities, was our Universe born with the conditions that allow life to arise?

That is the subject of this book.

AN INTRODUCTION TO FINE-TUNING

What do we mean by fine-tuning? Let's start simply by thinking about baking a cake (Figure 1). The first step might be to get your favourite cookbook and find a recipe – a list of instructions to go from raw ingredients to tasty cake. You combine the ingredients in order, stir and mix, bake for an hour, and finally turn out onto a cooling rack. You know that while the recipe says add two cups of flour, with a little bit more or a little bit less the cake should still turn out alright.

However, doubling the amount of flour, while keeping all the other ingredients the same, could end in baking disaster. And anything more than a pinch of salt would be very unpleasant. You could, of course, double all of the ingredients, cook for slightly longer, and end up with double the cake!

So, the cake recipe is somewhat fine-tuned. You can slightly vary the amount of each of the ingredients and end up with tasty cake. You can also scale the amounts of *all* of the ingredients up or down, and if you adjust the cooking time appropriately, you'll be fine. But deviate too far and you'll probably make an inedible mess. Certainly, if you throw ingredients in at random, and scramble the order of mixing and baking, the chances of something edible emerging are rather small.

So, are the conditions for life fine-tuned?

Let's consider a simple example that we'll come back to later. Everything that you can see is composed of atoms, tiny balls of positive charge surrounded by orbiting electrons. And each electron has exactly the same mass. Just how different would the Universe be if it had been born with electrons with twice the mass? In this hypothetical universe, the electron orbits would be different, changing the size of the atoms, and hence the molecules from which they are built. Perhaps this new mass makes little difference, allowing beings like us to exist. But what if the electron mass had been a million or a billion times larger? With such different atomic and molecular physics, could complex life forms exist? Clearly, we can consider an infinite variety of universes, each with a differing electron mass, and the core question of fine-tuning is what fraction of these could support complex life.

Before continuing, there is a potential confusion with the term *fine-tuning* that we should address. To a physicist, 'fine-tuning' implies

FIGURE 2 A radio set can receive a wide range of frequencies, but only a precisely positioned dial will allow you to enjoy the Norfolk Nights on Radio Norwich[2]. 'Fine-tuning' is a term borrowed from physics, and refers to the contrast between a wide range of possibilities and a narrow range of a particular outcome or phenomenon.

that there is a sensitivity of an outcome to some input parameters or assumptions. Just like baking a cake, if an experiment produces some spectacular result only for a particular, precise set-up, the experiment is said to be *fine-tuned* with respect to the result. 'Fine-tuning for life' is a type of physics fine-tuning, where the outcome is life.

'Fine-tuning' is a metaphor, one that brings to mind an old radio set with dials that must be delicately set in order to listen to Norfolk Nights on Radio Norwich (Figure 2). This metaphor unfortunately involves a guiding hand that sets the dials, giving the impression that 'fine-tuned' means cleverly arranged or made for a purpose by a *fine-tuner*. Whether such a fine-tuner of our Universe exists or not, this is not the sense in which we use the term. 'Fine-tuning' is a technical term borrowed from physics, and refers to the contrast between a wide range of possibilities and a narrow range of a particular outcome or phenomenon. Similes and metaphors are perfectly acceptable in science – space expands like an inflating balloon, for example – as long as we remember what they represent.

So there's a difference between asking 'is the Universe fine-tuned for life?' in the physics sense, and 'was the Universe fine-tuned for life by a creator?'

---

[2] Home of Alan Partridge, superb comic creation of Steve Coogan.

A Sunny Day and a Conversation

Introducing tricky topics is never easy – if it were, then they wouldn't be tricky. So we look for inspiration from the birth of the scientific revolution, when Galileo faced exactly this problem when trying to promote the radical idea that we should remove the Earth from the centre of the Universe, and suggesting instead that the planets orbit the Sun. Of course, Galileo also faced the problem of conflict with the academic establishment and the Church, which could have hefty consequences in the seventeenth century.

Galileo's solution was not to write a monologue, unambiguously stating his case and publishing in an academic journal, as a scientist would do today. To present the competing 'World Systems', Galileo wrote a dialogue, where three protagonists, Salviati, Sagredo and Simplicio, argue the merits of rearranging the Solar System. Such a dialogue is reminiscent of discussions in academia, or at the pub. Or both.

In the following, we want to introduce the core concept of this book to you, namely the question of whether the Universe is fine-tuned to allow life to flourish. Some may think this is a rather empty question, but once we realize that we don't quite know why the Universe is the way it is, then the question 'what if things had been different?' becomes extremely interesting, and leads to some rather surprising conclusions.

Our dialogue will set the scene for the chapters to come, examining life and liveability by delving into our understanding of the very fundamental nature of the Universe. However, a dialogue can be hard work (reading a play of Shakespeare is a lot harder than seeing it performed) and forthcoming chapters will revert to a more typical writing style.

Of course, modern 'management-speak' has got rid of dialogues, discussions, debates and diatribes, and so to please middle management everywhere, we present an action-oriented brainstorming conversation to identify additionalities[3] pertaining to the fine-tuning of the Universe for life.

Narrator: Our scene is set amongst Sydney's sandy beaches and rocky cliffs. While the parts of Sydney that the tourists don't see, including the arterial highways and apartment blocks, are filled to bursting point, there are many beautiful and serene pockets where one can sit and think about life. Our story starts in one such corner, on a gloriously sunny day, with two cosmologists thinking about the Universe.

Geraint: It's an amazing time in astronomy. For decades, we've known that there are billions of stars in our own galaxy, and billions of galaxies in the Universe. Thanks to the Kepler space mission, we now know that most stars have planets. Lots of planets could mean lots of life!

Luke: Yes, there are lots of planets, but that does not necessarily mean that there is lots of life. And even if life were common, we would expect much of it to be little higher than pond scum. Boba Fetts and Spocks may be very few and far between.

Geraint: But life arose here! And if the laws of physics are the same everywhere in the Universe, then shouldn't we expect the prospects for life to be similar?

Luke: It takes more than the same physics. Obviously, if you're going to make carbon-based, oxygen-breathing, star-powered life, then you'll need some carbon, some oxygen, and the occasional star.

But we don't know how life first arose. We have some clues about how it could happen, but no one knows the chemical reactions that connect the warm little pond of chemicals to a living cell. Still, there are places that look obviously worse than Earth.

Geraint: I guess we only have to look at the distant lumps of rock in our own Solar System. Pluto is frozen, and any life there, deprived of any significant heating by the Sun, would proceed at a snail's pace.

---

[3] This phrase was repeated many times at a 'scientists should be more entrepreneurial' seminar we attended. We have no idea what it means.

Luke: Right. Life needs the right kind of environment. But the laws of physics also play a key role.

Geraint: How so?

Luke: Well, in a few ways. The *laws of physics* have several key parts. Firstly, there are the building blocks of the Universe, the stuff. Then there are the ways that these building blocks can interact, which are the fundamental forces. And the laws of physics also presuppose the stage, the space and time in which the building blocks exist and interact.

Geraint: OK. This is physics for beginners: particles, atoms, molecules, gravity, magnetism, light and radioactivity. The rulebook for how the Universe behaves.

Luke: Exactly. We are the result of the action of the laws of physics over the history of the Universe. It is these laws that power the Sun, forge the elements, build the planets, form the molecules, and drive the chemistry of life.

So now we can ask: What if? What if the laws of physics were different? What if the building blocks, atoms and molecules, had different masses? What if electricity and magnetism were stronger, or gravity repulsive? What if elements were more radioactive? Or there was no radioactivity at all? What if we messed about with the stage, playing around with the very space and time underlying the cosmos? What would change in the Universe? And what would it mean for life?

Geraint: But isn't that a rather silly question to ask? What's the point of playing 'what if' games?

Luke: Human curiosity, for a start. Life seems so contingent, so full of possibility. There are so many ways that things could have turned out: if only I'd caught that bus, that falling vase, that ball or that big break in Hollywood. The twists and turns of history have inspired academic essays with titles such as 'If Louis XVI Had Had an Atom of Firmness' and 'Socrates Dies at Delium, 424 BC', several shelves of novels that explore the coulds, woulds and mights of

Hitler winning WWII, and a hundred thousand (or so) forum posts at alternatehistory.com and counter-factual.net.

In science, we play 'what if' games for a few reasons. We want to know which of our competing theories is the best. We compare Albert Einstein's theory of gravity with Isaac Newton's theory, calculating which gives the most accurate description of the Universe we see around us. Part of that comparison is asking: what would the Universe be like if Newton's theory was true? What would we observe if Einstein got it right?

Also, even our best and deepest physical theories have loose ends. There are numbers in the equations that the theory cannot predict. We just have to measure them. They are called the *constants of nature*. Why do they have the value that we measure? If that question has an answer, it must go beyond our current theories. Perhaps we can get a clue from asking 'what if these constants were different?'

Geraint: Why think that they could be different? In other words, why think that these other universes are possible?

Luke: We don't know whether they're possible – that's what we want to learn from a deeper, simpler, more unified law of nature. Perhaps they are mathematical constants, and cannot be changed without replacing the entire theory. Perhaps they aren't constants at all, but vary from place to place.

Geraint: Even if we did play with the laws of physics, how different could the Universe possibly be?

Luke: Well, you might suppose that because life is so versatile, any old universe would manage to make *something* living. Life has pulled itself together from the hodgepodge of chemical reactions in this Universe. Perhaps any old chemical rulebook will do.

Or we could actually investigate these other universes. It's fun to think about what conditions would be like if we changed the laws of nature.[4]

---

[4]   Note that a cosmologist's view of 'fun' may be quite different from your own.

Geraint: Hmmm, OK.

Luke: The surprising thing, discovered by the scientists who did the necessary calculations, is that messing about with the laws of physics radically alters the workings of the Universe. Many universes are inhospitable for life, even completely sterile. Ruining a universe is easy.

Geraint: Well, that would seem to make our Universe a rather happy coincidence. How did all the right pieces come to exist in our Universe?

Luke: Exactly! That is the fine-tuning problem. Why does our Universe have a mix of fundamental particles and laws that allows us to be here to ask questions at all? The fine-tuning of the Universe for life is the realization that if the laws of physics were different, even just by a little bit, life would not exist.

Geraint: So, what's the solution?

Luke: Well, what do we do when we face something seemingly unlikely? Maybe it's just something unlikely – end of story. Maybe it isn't as unlikely as we think. Maybe it's like the lottery – a winning ticket isn't too unlikely because lots of people buy different tickets.

That last idea, applied to the fine-tuning of the Universe for life, is rather ambitious. It supposes that a universe that is right for life exists because there are untold multitudes of universes with different properties. In the cosmic lottery, we got lucky.

Geraint: Sounds like science fiction.

Luke: Some think so. Others, seeing the lack of plausible ideas for explaining the values of the constants of nature, take the idea seriously.

Geraint: And us?

Luke: We're writing a book about it.

REVISING THE BASICS

Before we can start the journey of this book, we need to prepare by asking a few seemingly simple questions.

Question 1: What Is Life?

We're going to be talking a lot about life. We'd like to start with a definition, but this immediately lands us in trouble. Life has proven to be a very difficult concept to define precisely. We can all see the difference between the kind of thing a rabbit is and the kind of thing a rock is. A rabbit can see a fox approaching and run into its burrow; a rock might be pushed into a hole by the wind, but that's a very different kind of reaction. Is life defined by its ability to respond to the outside world? Rocks respond to the wind. But the rabbit reacts to the information that 'a fox is coming', even if it doesn't consciously think that thought. Is that what defines life?

Or is it the ability to reproduce? Rabbits famously make more rabbits; rocks can be crushed into a multitude of smaller rocks, but again that's a very different kind of thing. Rabbits make more rabbits via an internal rabbit-making recipe. The instructions for rabbit production are inside the rabbit, coded as information, and implemented via biological reproduction. Tweaking this biological code is what makes each generation, and each species, different.

And yet, suppose we met an alien race with which we could chat casually about the weather on Mars and what they've learned about the laws of nature. If an alien happened to mention that their species doesn't reproduce – perhaps they are sterile drones, descended from a long dead queen but able to live indefinitely – we wouldn't offend our guests by blurting out: 'Oh, I'm sorry ... I thought you were alive.'

Living creatures need to draw energy from their environment and put it to use. So is this *metabolism* the defining characteristic of life? More generally, life seems to have the ability to maintain an internal, ordered state against a changing environment. Life forms grow and flourish; they don't simply erode and decay.

One of the problems with crafting a definition for life is the hard cases, the borderlines between living and non-living. Is a virus a life form, even though it doesn't reproduce by cell division? What about *prions*, which are little more than badly formed protein molecules, but are responsible for mad-cow disease? Viruses and prions replicate by hijacking the machinery of a healthy cell, but is this life?

Computers and robots can respond to information about their environment. Are they alive? Crystals can form, grow, and create structure. Are they living, even though they don't do these things in accordance with an internal code, like DNA (deoxyribonucleic acid) in our cells?

Our discussion will touch on even more woolly questions about life. We will be concerned with the conditions under which life *forms*, and how common such conditions are in our Universe and beyond. It would be wonderful if, like our cake mix, we could simply provide a recipe for life:

> **1** star
> **1** planet surface (not too hot or too cold)
> Sprinkle your planet's surface with
>> **10** parts water
>> **5** parts carbon
>> **3** parts oxygen
>> A pinch of hydrogen, nitrogen, calcium, phosphorus, potassium, sulphur, five spice, olive oil, a squeeze of lemon (to taste).
> Bake using the residual heat of the early stages of the planet's formation. When the crust is firm, grill in starlight for a billion years, continually moistening with the water from colliding comets, until firm to the touch.
> Stir with meteorites and volcanos.
> Serve at room temperature (with garnish).

Unfortunately, we have only clues as to the sequence of events by which life formed on Earth. This is an extraordinarily difficult scientific problem, for three reasons. Firstly, life – even a single, 'simple' cell – is a miracle of complexity. Every cell in your body, for example, has molecular machines for moving itself, tagging and transporting molecules, processing food, defending against invaders, DNA duplication and repair, producing proteins and receiving and processing outside signals. On top of all that, this entire machine can tear itself in half and produce a complete working copy in about 20 minutes. A modern computer is pretty great, but it can't do that.

Secondly, the study of the origin of life is a *forensic* science. Like a detective gathering clues, scientists are trying to piece together a microscopic event, but are four billion years late to a crime scene that is the size of the Earth, and constantly moulded by water, wind, shifting tectonic plates, volcanos, sunlight and the occasional catastrophic meteorite impact.

Thirdly, and even worse, the origin of life could be an extremely rare event, even given the 'right' conditions. The process by which life forms could be so unlikely that it has only happened once in the galaxy, or worse. This makes the scientists' job much harder, as they may be looking for a singular set of circumstances. Which statistical fluke was responsible for life as we know it?[5]

Should we just stop here? If we don't know the conditions for life, how do we know how those conditions change with the physics of the universe?

Let's dive into an example, previewing later chapters. Our Universe appears to contain a form of energy that has *anti-gravity*. We know this from its effect on the expansion of the Universe, but we don't know what it is. To reflect this ignorance, we have given it the name *dark energy*: a nicely mysterious name that ensures that cosmologists pique the media's interest.

Dark energy could be a number of things, including something called *vacuum energy*, that is, the energy present in empty space even when there are no particles. Our best theory of the structure of matter

---

[5] Doesn't this make it unlikely that life formed by natural processes? To calculate the probability that life forms *at all* in the Universe by natural processes, we would need to know the size of the Universe. How many opportunities are there for this unlikely event to happen? We don't know the size of the Universe, so we don't know how to do this calculation. There is no reason to believe that the size of the *observable* Universe (the part of the Universe from which light has had time to reach us here) is any indication of the size of the *whole* Universe.

tells us that each fundamental type of matter will contribute to this vacuum energy, either positively or negatively. Alarmingly, the typical size of these contributions is larger than the amount of dark energy in our Universe by a factor of 1 followed by 120 zeros, or in scientific notation 10^{120}.

What would happen if the amount of dark energy in our Universe were, say, a trillion (10^{12}) times larger? This sounds like a big increase, but it is a pittance compared to 10^{120}. In that universe, the expansion of space would be so rapid that no galaxies, stars or planets would form. The universe would contain a thin soup of hydrogen and helium. At most, these particles might occasionally bounce off each other, and head back out into space for another trillion years of lonely isolation.

We may not know exactly what life is, or exactly how life forms, but we know that life isn't *that*. Such a universe would be fantastically simple, since matter would never get together in large enough numbers to make anything more complicated than a hydrogen molecule. Because gravity won't make matter collapse into galaxies or stars or planets or *anything*, physics is easy. Too easy. Too simple for anything like life.

At this point, people often play the science fiction card, and retort that such a simple universe could contain life not as we know it, life so extraordinary and bizarre that our puny human minds could not even conceive of its existence. But the important word here is *fiction*. Any genesis of life we consider must be based in science, not science fiction. Any universe in which life can arise must provide the conditions for the storage and processing of information; a thin soup of only hydrogen and helium simply does not provide this.

Let's continue thinking about simple vs. complex universes with an illustration. Suppose we're trying to invent a new board game. It will be a bit like chess, but with slightly different rules. As a first attempt, we'll make one small change to the rules: instead of stating that the only piece that can jump over other pieces is the *knight*, our new game says that the only piece that

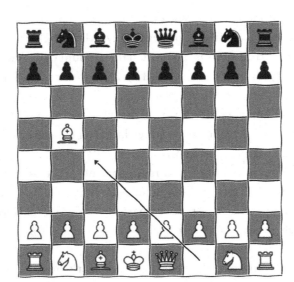

FIGURE 3 How to checkmate in schmess.

can jump over other pieces is the *bishop*. Instead of chess, we've invented *Shmess*.

Is shmess an interesting game? Wait a minute … we haven't defined the term *interesting*. How can we decide whether a game is interesting if we don't know exactly what that term means, or if different people find different things interesting?

In the end, it doesn't much matter. Part of what makes chess interesting to its admirers is the intricacy of its strategy. Chess strategy textbooks can be hundreds of pages long, and Grandmasters spend a lifetime mastering the game. If, on the other hand, we were to write *An Introduction to Strategy in Shmess*, it would need just two sentences: 'White moves her bishop from f1 to b5. Checkmate.[6]' That's it. The game is over before Black has his first move (Figure 3).

We don't need a precise definition of *interesting* to conclude that a game in which one player always wins and the other player always

---

[6] Technically, it's 'shmeckmate'. But you're just learning so we'll keep it simple.

does nothing is not an interesting game. The game is too simple. We know what would happen in a game of shmess, and we know that none of those things is interesting.

Let's expand the example. Suppose you've been inventing new board games all afternoon. You've tried a thousand different sets of rules, and all but two are as boring as shmess. Now, we could argue about which definition of interesting is *really* the right one, and whether these two games are *really* interesting. But the big story here is how rare interesting games are in the set of possible games – a conclusion that we can reach without precisely defining *interesting*.

The reason is that, in order to conclude that most games are not interesting, we don't need to decide the borderline cases. We only need to be able to identify obviously non-interesting games.

Similarly, all we need for an investigation of fine-tuning is to be able to identify examples of obviously non-living things. If a universe is simple enough, we can safely conclude that nothing as complex as life could form.

There are hypothetical universes whose laws and constants of nature, while not a definitive death sentence for all life forms, are certainly a dramatic step in the wrong direction. For example, a super-villain with his hand on the cosmic dials could crumble all your atoms into a pile of hydrogen. While it is conceivable that some form of life could exist somewhere in such a universe, a call to your favourite superhero would probably be wise.

As a result, we needn't worry too much about a precise definition of life. A typical dictionary definition will do: life is characterized by the capacity to grow, metabolize, actively resist outside disturbance, and reproduce.

Question 2: What Is the Anthropic Principle?

Scientists and philosophers have debated the extent and implications of the fine-tuning of the Universe for life for several decades. Debates have also raged among interested laypersons. Sooner or later, someone will mention the *Anthropic Principle*.

Discussion of the anthropic principle is clouded by its many, contradictory definitions. We need to clear up this mess, and will do so by tracing the origin of the confusion.

Australian-born cosmologist Brandon Carter introduced the term in a now famous talk in Warsaw in 1973. Here is Carter's *Weak Anthropic Principle (WAP)*:

> We must be prepared to take account of the fact that our location in the universe is necessarily privileged to the extent of being compatible with our existence as observers.

One version of the history of science tells of humankind's gradual realization that they are not the special, unique, all-important centre of the Universe. Medieval mythology arrogantly presumed that the cosmos revolved around us, only to be overthrown by Copernicus and Galileo. We are not at the centre of the Solar System, much less at the centre of the Universe. To such a view, Carter's principle seems obsolete.

However, history tells a different story. It was not the medievals who placed the Earth at the centre of the Universe but the ancients. Specifically, Aristotle's cosmology of the fourth century BC consisted of around 50 transparent spheres rotating around the Earth. The stars and planets are made of different stuff – celestial aether – that is perfect and incorruptible. By contrast, Earth is made of, well, earth. While it is the very nature of aether to maintain perfect circular motion, earth's weight and imperfection causes it to sink. Our home planet isn't at the centre; it's at the *bottom*! It's where the crud of the Universe collects.

Aristotle had his reasons for such a system, and they do not involve human arrogance[7]. Rather, they are empirical. When you

---

[7] We would do well to remember that, while the Hebrew Scriptures place humankind near the pinnacle of creation, the Greek and Babylonian stories do not. The Babylonian Enuma Elish tells of a primordial battle between the chaos monsters

jump, you land in the same place. You don't land 500 metres to the west, which proved to the ancients that it is the heavens that are moving and not the Earth. (Only when one understands Galileo's relativity of motion can this argument be countered.) But motion on Earth doesn't last. If your horse stops pulling its cart, it quickly comes to rest. If you drop anything made of earthly matter, it falls back towards its natural place in the scheme of things, and comes to rest. So the heavens, with their perpetual, perfect, circular motion, must be made of different stuff, and kept in motion by the *Primum Mobile*, the outermost and greatest of the spheres.

It is preposterous, then, for the ancients and medievals to join Copernicus in moving the Earth out into the heavens. This is not because it demotes us from our throne at the centre. Quite the opposite – it puts us in too high and lofty company. We don't belong out there among the perfect spheres. Earthly stuff doesn't move like heavenly stuff. And how could we possibly place the Sun – the perfect source of light and life – at the bottom of the Universe? What had it done to deserve a seat of such dishonour?

New physics, and in particular a new understanding of matter and motion, was needed. The revolution was glimpsed by Galileo and completed by Newton. All objects remain in a state of constant motion unless acted on by a force. The planets move in circular orbits due to the gravitational force of the Sun; otherwise, they move largely unimpeded through practically empty space. Earthly things come to

Marduk and Taimut, the leaders of the competing factions of gods. Marduk triumphs, and rips the corpse of Tiamut into two halves from which he fashions the Earth and skies. Kingu, a rebel god who incited the war, is destroyed so that from his blood Marduk can create:

> . . . a savage, 'Man' shall be his name.
> Verily, savage-man I will create.
> He shall be charged with the service of the gods
> That they might be at ease!

The epic ends with a hall of feasting gods chanting the 50 kingly names of Marduk. Whatever inspired the story that humankind exists to be the slaves of the reigning chaos monsters, it wasn't human self-importance. Greek mythology has a similarly low view of humankind's place in the grand scheme of things.

rest because of other forces – friction, air resistance, contact forces. In this way, we can explain earthly and planetary motion in terms of the same principles and the same matter.

Modern astronomy shows that we are not even at the centre of our galaxy. (It's probably just as well – the centre of our galaxy hosts a black hole that is a million times heavier than the Sun.) We are the third planet around a typical star in an average-sized galaxy in a universe with planets, stars and galaxies in every direction. Not only are we not at the centre of the Universe, there is no centre.

So just what does Carter mean when he says that our location must be privileged?

Consider a simple example. We usually take air for granted, but the density of the air you are breathing is 10^{27} times the average density of material in the Universe. Places in the Universe with a density at least as large as the air in a room are cosmically rare. Why would you, a human being, find yourself in such a rare location?

The answer is not difficult to discover. Humans are the result of billions of years of evolution, built out of a myriad of complex molecules and structures. This process requires an environment rich in chemicals, and dense enough for efficient chemical reactions. Humans should not be surprised to find themselves in such an environment, even if it is rare.

In fact, any other intelligent beings in our Universe that are questioning their existence will probably find themselves in such *privileged* environments.

We can take this argument further. When Carter says *location*, he means not just in *space* but also in *time*. We expect life to be more likely to arise not just in certain places but also at certain times.

The early Universe consisted of mostly hydrogen and helium, with virtually none of the elements for creating planets, trees and people. The Universe needs to create several generations of stars to produce large quantities of carbon, oxygen and other elements. As an intelligent being, you should not be surprised to find yourself in a Universe that is almost 14 billion years old, that has had sufficient time to create the material needed to create you. You exist at a *privileged* time.

WAP says: the Universe is not your experiment, to set up as you please and observe at your leisure. You are not Dr Frankenstein. You are the monster. You have awoken amidst the beakers, electrodes and dials of the machine that created you. *What* we observe may be affected by the fact *that* we observe at all.

Carter took this line of thinking one step further, introducing the *Strong Anthropic Principle* (SAP). It says:

> The Universe (and hence the fundamental parameters on which it depends) must be as to admit the creation of observers within it at some stage.

Simply put, WAP asks: why here? why now? SAP asks: why these physical laws and constants? WAP is about our place in space and time. SAP is about the properties of the Universe, such as the values of the constants of nature.

Carter's SAP is easily misunderstood; the source of most confusion is the word *must*. The sense is not logical or metaphysical, that is, that a universe without observers is impossible. Neither is it causal, as if we made the Universe. Rather, this *must* is consequential, as in 'there is frost on the ground, so it must be cold outside'. *Given that we exist*, the Universe (and its laws) must allow observers.

Carter's WAP and SAP are about what follows from our existence as observers, and so cannot explain why observers exist at all. These principles are mere tautologies, unable to explain anything. However, similar tautologies play a role in scientific explanations of the world. A telescope can see only objects that are bright enough for it to see. Only people who respond to the survey will be surveyed. The organisms best able to survive are more likely to survive. These are not the whole explanation of some phenomenon – natural selection, for example, involves more than survival of the survivors. But they can be important.

Here's where the confusion starts: later writers have not followed Carter. In 1986, two well-known physicists, John Barrow and Frank Tipler, published an influential book titled *The Anthropic Cosmological Principle*. They delved into questions about the existence of intelligent life and its implications for the laws of nature. It is a wonderful book, but it less-than-subtly redefines the weak and strong anthropic principles, causing considerable confusion. According to Barrow and Tipler (p. 16), the Weak Anthropic Principle states:

> The observed values of all physical and cosmological quantities are not equally probable but they take on values restricted by the requirement that there exist sites where carbon-based life can evolve and by the requirement that the Universe be old enough for it to have already done so.

This is, in fact, a combination of Carter's weak and strong principles. It refers to 'all physical and cosmological quantities', including space and time (Carter's weak principle) and the constants of nature (Carter's strong principle). It is, we contend, reasonable to combine the two, but the result should be simply called the *anthropic principle*.

How, then, do Barrow and Tipler define the Strong Anthropic Principle?

> The Universe must have those properties which allow life to develop within it at some stage in its history (1986, p. 21).

This is where things get interesting. They offer several alternative interpretations of this statement, including:

1. There exists one possible Universe 'designed' with the goal of generating and sustaining 'observers'.
2. Observers are necessary to bring the Universe into being.
3. An ensemble of other different universes is necessary for the existence of our Universe.

We're a long way from Carter's SAP. The 'must' in Barrow and Tipler's Strong Anthropic Principle is taken to imply that intelligent life is somehow central to the very being of the Universe, even suggesting we made it!

With this redefinition, the Strong Anthropic Principle becomes quasi-metaphysical, making philosophers thoughtful and scientists suspicious.

This redefinition is unwise. WAP and SAP are supposed to be stronger and weaker versions of the *same kind* of principle. Carter's principles are: the same idea is applied narrowly to space and time (WAP) and more widely to the constants of nature (SAP). However, Barrow and Tipler's motley company of ideas – from circular to speculative – march under the same 'anthropic' banner. This has tended to give them all an undeservedly controversial air. Even Carter's utterly obvious WAP is viewed with suspicion because of its dubious namesakes.

We will leave the anthropic principle for now; it will pop up here and there throughout the book. If you can't wait, and have plenty of time on your hands, typing 'anthropic principle' into your favourite search engine will provide hours of entertainment, though significantly less enlightenment.

Question 3: What Is Science?

We will be tiptoeing around the fringes of science. We need to know when we've wandered too far, straying into speculation, metaphysics, or worse.

Being scientists, our view of the scientific enterprise will be from the inside. We are most familiar with our field and our colleagues and our projects, and must step back to generalize about science and scientists and the scientific method. In particular, transcending our time and culture to paint an authentic portrait of the history of science is no triviality. Inevitably, our account of *the* scientific method will be coloured by the goings-on in building H90 of the University of Sydney.

In fact, this 'scientific method' is a bit of a myth. It sounds like scientists have a little book with stern rules about asking questions, defining hypotheses, and performing experiments to decide if your ideas are rejected by the ugly facts of nature, or live to fight another day. In reality, the process of science is learned on the job, meaning it is a rather messy process in practice.

Our focus will be the field of physics, as this is the most familiar to us and the most relevant to fine-tuning. Physicists are often lumped into two camps; theorists who try to construct the mathematical rules of the workings of the Universe, and experimenters who investigate how the Universe actually behaves. In reality, the distinction between theorist and experimenter is not perfectly sharp, with many people having a foot in both camps, but we'll stick with the distinction for now. Let's start by looking at the role of the experimenters.

The Experimenters

Well, *someone*'s got to actually look at the Universe.

Experimenters come in various shapes and sizes. In astronomy, for example, we are typically passive observers of the Universe we see around us. We can't test our ideas about stars by making one in the lab.

More typically, we picture an experimenter in a lab, surrounded by instruments, chemicals and brains in vats. These experimenters tinker with nature, sending electrons one way or another, or placing crystals into super-strong magnetic fields just to see what happens. In the physical sciences, however, it is not sufficient to simply describe your observations in words (although this can be important). We need to get *quantitative* – we need numbers. This recording of the properties of things, especially how they change as an experiment is tweaked, is a vital part of science.

Consider a simple question: what colour is the sky? These words are being typed on an airplane flying between Sydney and Melbourne, and outside there is the lovely view of the winter sky over Australia.

The sky could be described as being light blue in colour (ignoring the scattered clouds below), but it is somewhat darker above, and becomes lighter towards the horizon.

A nice description, but how does this patch of sky compare to that observed somewhere else on the planet? To meaningfully compare, we need to measure physical properties of the sky at each location. We can't just exchange impressions; we need numbers.

Light is a form of energy, and we know that it comes in a range of wavelengths. So we could build a device to measure how much energy is deposited in my eye by the light received from the sky. We can also measure how much energy is deposited as we change the wavelength of the light. Such devices exist and are known as spectrographs, as they split the light they receive into a spectrum of colours. All Pink Floyd fans know that a glass prism can split a beam of white light into a rainbow of colours.

We can measure how the sky looks at different wavelengths, and also as we look further above the horizon. With these measurements, we can compare the sky at different locations on Earth.

Unfortunately, the real world is messy. Equipment and detectors are never perfect, and any data we record will come with an associated uncertainty, something that is often referred to as an error, although it does not mean that something is wrong. A better word is noise[8].

For example, suppose we leave our detector out in the sunshine and it collects 10,784.3 joules of energy with an uncertainty of 0.1 joules. We can say that the actual amount of energy falling on the detector is likely to be in the range 10,784.2 to 10,784.4, very likely to be in the range 10,784.0 to 10,784.6, and that it is extremely unlikely that only 100 joules or 100,000 joules arrived.

---

[8] Our next project will be to write a book about noise and uncertainty in science. In essence, these are the most important things in all of science, but are poorly understood by those not in the field. We dream of the day when it is routine for the media to report uncertainties on a measurement.

Scientists have had to deal with the uncertainties in their measurements for a long time, and have robust mathematical approaches to deal with them. While these methods, known as Bayesian statistics, are widely known, they are not always applied. Why science operates this way is a topic for another book!

More than just looking, experimenters measure, determining the properties of the world around us. But this cataloguing is just one side of science.

The Theorists

What do we do with a mountain of measurements? We can search for patterns and trends, hoping to look behind the data and see the inner workings of the Universe. In physics, theorists seek the mathematical laws by which the machinery of nature operates.

While mathematics has long been appreciated for its beauty and usefulness, its crucial role in physics is a relatively recent discovery. Students in medieval universities were first taught critical thinking via the *trivium*: grammar, logic and rhetoric. They were then taught the *quadrivium*: arithmetic, geometry, music and astronomy. Music might seem out of place, but students learned not performance or composition, but the mathematical theory of harmonics and proportions.

Similarly, astronomy was viewed in the tradition of Aristotle as a 'middle science', living between abstract mathematics and empirical (but largely not quantitative) physics. We could theorize in mathematical terms about the geometry of the heavens, but it seemed inconceivable that such symmetry could be found down here.

Rene Descartes, in the early seventeenth century, most clearly envisioned and championed the idea that all of physics could be as mathematical as astronomy. Descartes had a vision of 'unifying all sciences of quantity under mathematics'.[9]

---

[9] Williams, 1978 (p. 16).

The first task was to unify celestial and earthly mechanics, astronomy and physics. Descartes' own attempts were unsuccessful, in part because he thought that a vacuum was physically impossible. Building on the important work of a great many scientists, including Kepler and Galileo, it was the towering genius of Isaac Newton who first fulfilled Descartes' vision. Theoretical physicists follow in his footsteps.

The theorist's primary tool is a *model*, built not of Lego® blocks but of mathematics. Returning to the blueness of the sky, our model needs a few components. We need to know about the source of the light we receive from the sky – sunlight – and, in particular, its distribution of energy as a function of wavelength. We need the properties of the atmosphere: its various gases with their molecular structures, and how the atmosphere changes with altitude, being warm and dense near the ground, colder and rarefied up high.

The model would also have to consider how light moves through the sea of molecules that make up the atmosphere. Does light sail through or does it scatter off the molecules? And if it scatters, how does this change with the wavelength of the light?

At all stages, the theorist will call on what is known, such as the molecular make-up of the atmosphere and the pressure and temperature at various heights. They may require new calculations, such as deducing how light scatters differently off nitrogen and oxygen molecules.

Four pieces make up the theorist's model. The physical material is represented by a mathematical *object*, that is, something like a set of numbers or a function or a field or a manifold that captures everything the model says about the system. For example, a collection of classical particles is represented by the position and velocity of each particle in space and time. For the gas in a room, it could be the temperature, pressure and density at each point. A variety of elaborate and sophisticated mathematical objects are available.

The second piece is the mathematical *form* of the equation. This encodes how the stuff moves, acts and reacts. With different equations, our classical particles might attract or repel, our fields might wiggle, or quantum things might ... do whatever quantum things do. Physics equations are typically *dynamical* – they tell us how a system changes over time.

The third part is a set of *constants*: plain old numbers. They might tell us how strongly two particles repel each other, or how heavy they are. These constants are – by definition – not able to be calculated by the equation. They must be measured.

The fourth part is the *scenario* to which the equation is applied. Mathematicians call these 'initial conditions'[10] – the equation tells us how the stuff *would* act in a certain situation (move this way, bounce that way, swerve over here), so we need more information to specify how it actually acted. For example, given Newton's theory of gravity, we can investigate the Solar System, a cluster of stars or even the whole galaxy. Given a physical theory describing how electrons flow through a wire, we can investigate all kinds of electrical devices.

Here are the four pieces: the stuff, the dynamics (encoded in an equation), the constants and the scenario. Figure 4 shows how the pieces of a theoretical model come together to predict the orbit of a small particle (m_1) around a larger one (m_2).

The Crunch

Now comes the crunch: compare your theoretical prediction with observation. This is the essence of science. Actually, strictly speaking, this *is* science! Our models are mixtures of well-tested theories, reasonable assumptions and guesses; as Richard Feynman noted, 'it is not unscientific to make a guess.'[11] Science happens when we ask the Universe whether we guessed right. Otherwise, the experimenter is doing little more than stamp collecting, and the theorist is just playing with numbers!

---

[10] Or, more generally, boundary conditions.   [11] Feynman (1965, p. 165).

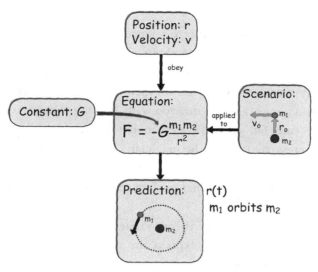

FIGURE 4 We use the example of applying Newton's theory of gravity to predict a planetary orbit. Four pieces must converge. A *mathematical object* represents the state of the system, in this case the position and velocity of the particles. The *equation* relates the state of the system to how it is changing in time. The equation requires a *constant*, G, that controls the strength of gravity. And we apply this general equation to a particular *scenario*, in this case two masses arranged and moving as shown. The result is a prediction: the planet (m_1) will orbit the star (m_2).

Scientists use probability theory to tell them how well a prediction matches the observed universe, and whether the data favour one physical theory over another. The poorly performing are relegated to the minor leagues, while the better ones are reemployed, and continue to be used until they no longer provide an accurate description of the world around us, and new insight is required to construct a better theory.

In this way, science finds increasingly accurate mathematical theories that predict what we observe in the natural world.

BACK TO FINE-TUNING

So, where does fine-tuning fit into science?

We said that science was about comparing your theory with data. However, there's a bit more to it. We prefer theories that aren't ad hoc, jerry-rigged, or too flexible for their own good. In general, if you have 10 data points, and an equation with 10 *free parameters*,[12] then your model can't fail – it will always match the data. Successful, sure, but not impressive. It's like the magician correctly guessing the card you selected from the deck ... on the 43rd attempt. Much more impressive are the theories that explain mountains of data with few moving parts.

Similarly suspicious are theories that need very precise values of free parameters in order to explain the data. To understand why, consider this little tale.

A bank vault is robbed. The armoured door was opened without force; the robbers used the access code. The police arrive on the scene.

DREBIN: Maybe they guessed the code.

HOCKEN: No way, Frank. There are a trillion combinations. The system shows that they entered the code correctly on the first attempt. Surely the odds against that are astronomical.

DREBIN: But it's still possible, right?

Here is one way to see the problem with Drebin's theory: it is composed of trillions of sub-theories. There's the sub-theory in which the robbers turn up and punch in 0000-0000-0000. There's the sub-theory in which they punch in 0000-0000-0001. And another with 0000-0000-0002. And so on.

On Drebin's 'they just guessed' hypothesis, each of these sub-theories is intrinsically equally likely. And yet, only one explains the fact that they got into the safe – the one in which they punch in the

---

[12] A free parameter is a number that can be adjusted to make a model fit the data. When fitting a straight line to some data, we would use the equation $y = mx + c$, the numbers m and c are free parameters; m is the slope, and c is the intercept, and for different values we get different straight lines.

correct code. This makes Drebin's theory suspicious. The 'which code they guessed' part of the theory must be *fine-tuned*. This does not sink the theory, but it does leave the door open for an alternative theory to better explain the data.

This is what the physicist means by 'fine-tuned' – *a suspiciously precise assumption.* (Precision is great in our data, but not in our assumptions.)

Physical theories, like physical systems, can be hierarchical – we build big ones out of small ones. The most fundamental laws we have describe the smallest building blocks of physics: electrons, quarks, photons and a host of other characters we'll meet in the next chapter. This is the domain of *particle physics*.

Similarly, the most all-encompassing scenario we can hope to model is the entire Universe. This is the domain of *cosmology*. Looking for the ultimate initial conditions sends us back to the beginning of time, and more of that in Chapter 5.

So, if the free parameters (constants or initial conditions) of particle physics and cosmology are suspiciously precise, then we have found fine-tuning at the deepest level of our understanding of the Universe.

The fine-tuning of the Universe for *life*, then, is fine-tuning applied specifically to the fact that this universe supports life forms. The claim is that small changes in the free parameters of the laws of nature as we know them have *dramatic, uncompensated* and *detrimental* effects on the ability of the Universe to support the complexity needed for physical life forms.

THE FUNDAMENTAL CONSTANTS OF NATURE

Let's take a closer look at some of these free parameters.

The electron is one of the fundamental particles of the Universe. Electron orbits around the nuclei of atoms dictate the processes of chemistry. With the appropriate experimental equipment, we can measure the mass of an individual electron: $9.109\,382\,15 \times 10^{-31}$ kg (and, with our most accurate equipment, we know this value has an uncertainty of 0.000 000 45 × 10^{-31} kg). If you measure the mass of any electron in the Universe, you get the same answer!

When we measure the mass of an object in kilograms, we are implicitly comparing it to a lump of platinum–iridium alloy held in uniform conditions at the International Bureau of Weights and Measures laboratories in the outer reaches of Paris. There is nothing special about this lump, and so nothing special about the kilogram. Nothing changes if we were to express the mass of the electron in pounds, long tons, grains or carats.

However, the mass of the electron relative to other particles in the Universe is important. Each member of the menagerie of fundamental particles comes with a mass, and while some are zero, many are just plain, unexplained numbers.

Here we can play our 'what if?' games. If we change the relative masses of the fundamental particles, what effect does this have on a complex, multi-cellular, balding primate sitting and typing on a planet orbiting a star? We'll see in later chapters that the existence of life depends critically upon particle masses. Universes with different mass ratios are often sterile.

Another fundamental aspect of the Universe is *force*. Pushes and pulls in everyday life come from friction, wind, springs, walls, gravity, motors, muscles and more. At the microscopic level, four forces are enough to model all known interactions between fundamental particles. They are gravity, electromagnetism, and the enigmatically named strong and weak nuclear forces.

Consider gravity. Newton described gravity with his famous 'inverse square' law: any two masses attract each other, with a force that decreases with the square of the distance. Einstein's General Theory of Relativity[13] is a more accurate and more difficult improvement on Newton's theory. In both theories, a quantity

---

[13] Small bugbear here, but some people talk about Einstein's Theory of General Relativity, not his General Theory of Relativity. The former is incorrect, as it is the theory that is general, not the relativity.

$$F = -G \frac{M_1 M_2}{r^2}$$

$$R_{\alpha\beta} - \frac{1}{2} R g_{\alpha\beta} = 8\pi G T_{\alpha\beta}$$

FIGURE 5 The gravitational force laws. Newtonian gravity on top and Einstein's version on the bottom. Don't sweat the details; just note that G appears in both equations, but cannot be calculated using those equations alone.

known as Newton's gravitational constant appears, which is usually given the symbol G and has a value of 6.67×10^{-11} m^3 kg^{-1} s^{-2} (Figure 5).

If the value of G were different, what would happen? We need to be a bit careful here. Suppose we've transported you to another universe and asked you to measure G. You'll need to calibrate your instruments to measure metres, seconds and kilograms. But wait ... that platinum–iridium lump is back in our Universe! Thankfully, changing G doesn't affect the elements, so we can (in principle!) make what we need. With some caesium 133, you can calibrate your clocks to measure seconds. Measuring the speed of light gives the metre: the distance light travels in 1/299, 792, 458 of a second. We can then construct a replica platinum–iridium lump to give us the kilogram. You can then measure G.

Nothing in Newton's or Einstein's theory tells us the value of G. We have to ask nature, measuring from experiment.[14]

In Newton's theory, if G were twice as large, the gravitational force between masses would be twice as large. In Einstein's deeper understanding of gravity, G measures how strongly mass and energy

---

[14] If you want to read a tale of dedication and experimental perseverance, you should find an account of how the eighteenth-century scientist Lord Henry Cavendish locked himself away with masses and fine strings to give the first accurate measurement of G!

distort the geometry of spacetime (more in Chapter 5). Changing the value of G affects just about everything in astrophysics, from the expansion of the Universe and the formation of galaxies to the size and stability of stars and planets.

Similar constants appear in all of the force laws, where they are called *coupling constants*. The only way we have of knowing the value of these constants is to measure them from nature.

In the next few chapters, we will play the 'what-if' game with particles and forces, revealing just how their properties influence the small and large workings of the Universe.

2   I'm Only Human!

We begin with a seemingly simple question: what are you made of? As you look at your hands, you can see skin, nails and hair, but underneath there is a whole lot more: bones, organs, and fluids of various colours. These are made of still smaller things: molecules and atoms, which are made of electrons, protons and neutrons.

In this chapter, we will explore the Universe from the inside out. We will investigate how the properties of the small influence the large, connecting the fundamental building blocks of the Universe to the machinery of life. We'll need to cut our way through bones, skin and cells, through layers of complexity into molecules and atoms, and eventually to the fundamental particles.

When we reach the bottom layers, we will find that all of the structures we see are made from a rather small number of building blocks. With this in hand, we will see how changing the properties of these basic pieces ripples up through the various levels of complexity, impacting the Universe on many scales, including its ability to support life.

WHAT MAKES A HUMAN?

This question has occupied humans for millennia. While 'what makes a human' in terms of consciousness remains a bit of a mystery,[1] over the last few hundred years science has revealed the building blocks of the human body in stunning detail.

Through scientific endeavour, we now know that a human body is a mass of interacting chemical signals, a complicated network of processes that allow us to eat, sleep, move about, and generally live.

---

[1] To some, this is a major understatement!

You're a very organized bag of chemicals, with lot of structure on a lot of levels.

With the naked eye (and with the aid of a sharp knife), we can see that the typical human is made from a number of key components, including a skeleton to support its weight, fibres of muscle and tendons for strength and movement, as well as an array of individual organs to process food and air into fuel. This is all conveniently wrapped in a semi-permeable skin, stopping everything spilling out onto the floor.

Anyone who has picked up a Human Biology 101 textbook knows the complicated processing that a morsel of food undergoes, starting at the mouth, through the gut where energy is extracted, and finishing in the sewer systems that make modern life bearable. As we look deeper, we see that the complexity of what makes a human continues to increase. The liver looks like a nondescript lump of flesh, but inside is sophisticated blood-filtering machinery. Blood appears to be just a red fluid, but when exposed to air, a complex cascade of chemical reactions creates a clot to stop the bleeding. Does looking deeper always uncover more complexity?

The simple, and rather amazing answer is no! Eventually, as we look deeper, as biology becomes chemistry and chemistry becomes physics, the complex gives way to the simple, until what makes a human can, in its rawest of terms, be scribbled on the back of a napkin. Let's start unwrapping the complexity by delving into what is revealed when we peer more deeply with microscopes.

THE MACHINERY OF LIFE

Every year, around the world, many millions of students arrive at university to embark on a new level of education, and part of the initiation is a visit to the campus bookstore, be it a physical or virtual location. The benefit of the physical bookstore is that you can browse the myriad of titles of courses that you are not doing and consider the mysteries of 'Cost Accounting', wonder what 'Boya Chinese' is, and think, what is the point of 'Modern Rhetoric'?

In recent years, physics education has replaced a broad range of specialized textbooks that cover the individual topics, such as classical mechanics, quantum theory, relativity and electromagnetism, with single, all encompassing volumes that weigh in at several kilos. If civilization falls, the cockroaches can use one of these books to relaunch modern physics out of the ashes.

Alongside the all-encompassing physics tomes, and with similar weight and girth, are textbooks that cover nothing but the biology of a cell. Why are they so massive?

Cells are the basic 'bits' of the body. The body contains a variety of cell types, each employed in differing roles. Some are well known, with red blood cells carrying oxygen to the muscles and carbon dioxide back to the lungs, or sperm and ovum that form the basis for new life.

Just as peering inside a human uncovered a wealth of differing structures – bones, organs, blood – that enable human existence, so each cell is a microcosm of activity. The players on this stage are complex *molecules*, structures built from individual atoms, which take part in elaborate interactions within the cell.

The complexity of these interactions is stunning. Each cell in your body, for example, has its own postal system. Though it is less than a tenth of a millimetre across, the cell cannot afford to lose track of the resources it has gathered from its surroundings. So, when a molecule needs to be sent from one end of the cell to another, it is stamped with an address, loaded onto a truck, transported down a highway, has its address verified, and then unloaded and put to work. Molecular machines do all this. Most impressively, the cell can reproduce, manufacturing a complete working replica in about 20 minutes.

If you are unfamiliar with the amazing microcosm within your cells, we urge you to find out about it right after finishing this book.[2]

---

[2] For those who do not want to read biology textbooks, we heartily recommend John Gribbin's *In Search of the Double Helix* (1985) for a less academic, but still absolutely fascinating, portrayal of cell biology.

We guarantee that you will be stunned by what is going inside your cells right now!

Just how do cells work? How do they do their various things? The secret has been revealed over the past century or so, starting with the realization that, on very small scales, everything is constructed of atoms. To understand the biology of life, we need to know what happens when atoms bind together to form molecules.

On Earth, lone atoms are rather rare; we are more familiar with molecules. When we drink a glass of water, it feels smooth and thirst-quenchingly continuous, but we know from our school science classes that it is actually an almost uncountable number of water molecules jiggling about. Each water molecule is essentially identical, consisting of two atoms of hydrogen stuck to one of oxygen (Figure 6).

Other molecules are available. Our atmosphere is dominated by nitrogen molecules, with two nitrogen atoms bound together. Also within our atmosphere we find carbon dioxide, one carbon and two oxygen atoms, which all humans breathe out, and which is responsible for so much angst due to its ability to blanket the Earth and warm the atmosphere. And, of course, 21 per cent of the atmosphere is composed

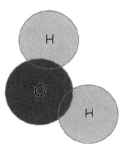

FIGURE 6 Consisting of one oxygen atom and two hydrogen atoms, water molecules appear to have a special role in the chemistry of life. Note that the atoms are to scale. The sizes of atoms are defined by the orbits of their electrons, not the mass of the nuclei. Even though oxygen is 16 times heavier than hydrogen, it is only 13 per cent larger because its more positively charged nucleus keeps the electrons close.

of the molecule we know, love and need, that of two oxygen atoms bound together.

All of these molecules are quite simple, with a few atoms bound together. Certain chemical elements are able to form long chains and loops of atoms, making complicated molecules. For example, floating amongst the water molecules in a cup of coffee are 2-furfurylthiol, 3-mercapto-3-methylbutylformate and others that give us the wonderful smell and taste of coffee.

And, of course, caffeine (1,3,7-trimethylpurine-2,6-dione), which is the right way to start the day. The complicated names are a necessary byproduct of the huge number of ways of combining atoms into molecules. Caffeine, for example, possesses a few tens of atoms wound and wrapped as in Figure 7. As we will see, this shape is vital for the way that molecules interact, including how caffeine gives you that little extra kick in the morning.

Perhaps the largest molecule you've heard of is DNA. This is a long-chain molecule, made from a twisted scaffold of molecular links. There are four key molecular blocks, known as guanine, adenine, thymine and cytosine, which bind the scaffold into the famous

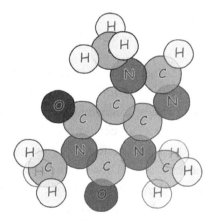

FIGURE 7 Caffeine, a complex molecule, continues to play an important role in human culture.

double helix. Like a string of letters, the order in which they appear along the DNA strand encodes information.

If we arranged the DNA in one of your cells into a straight line, it would be a few metres long, and just a few tens of atoms thick. How does it manage to pack into an individual cell? Here is another intriguing property of some molecules: they fold.

To understand this, let's go back to a simple molecule: water. The shape that a molecule takes is a battle between attraction and repulsion caused by the differing charges on the atomic nuclei and the electrons. When the individual atoms come together, the molecule settles down so there is an angle of 104.45 degrees between the two hydrogen atoms and the oxygen atom (Figure 6).

Larger molecules can fold in multiple ways, with the precise configuration depending on how they were formed. DNA, for example, folds into curls within curls, packing itself into the tiny volume available in the *nucleus* of a cell.

But what does DNA actually do? How does a big curled up molecule make you? The key word here is *proteins*. When most of us think of proteins, we think about something to do with meat, or diets, or maybe those massive, strangely decorated 'bulk-up' foods available in 'Health Food Stores'. To a molecular biologist, proteins are a kind of large molecule, some with more than half a million individual atoms. From proteins, cells can make more cells, tissues and organs.

How do cells make proteins? Here's the short version. Proteins are built out of smaller molecules called *amino acids*. An alphabet of 20 amino acids can be combined in cells, letter-by-letter, into a huge number of proteins. The proteins that a cell needs are coded in DNA, written in letters along the length of the strand. The relevant part of the DNA strand is unravelled and copied into a molecule called *messenger-RNA* or *mRNA*. The mRNA is used as a code by a factory called a *ribosome*. Groups of three DNA letters are read, and the corresponding amino acid grabbed from *transfer-RNA* (*tRNA*) and snapped onto the end of the growing protein chain.

As proteins roll off the ribosome production line, they wrap themselves into a more compact shape. Like DNA, proteins can pack their many atoms into a tiny volume. But not just any old volume, as the way that a protein curls up is essential to the way it interacts with other molecules, and hence essential to its function in the cell.

If you gently heat up a sample of protein, the extra energy causes the tightly bound protein molecules to unravel. If you then allow them to cool down again, the proteins curl themselves back into the previous shape. This is quite startling, as the long chain molecules have no memory of their initial shape, and it is only the push and pull of the electrons and nuclei that can guide them into position. We don't really know how they manage it; this 'protein-folding problem' remains one of the major outstanding problems of modern science.[3]

Why is the shape that a protein folds itself into important anyway? Because, for proteins, it's not what's on the inside that counts, it's what's on the outside! Once a protein has folded into its intricate shape, the atoms curled into the inside are effectively invisible to the outside world. But those atoms on the outer surface are, in management-speak, 'public-facing', and the shapes that these atoms present completely govern how the protein interacts with other proteins and molecular structures. These atomic chunks can lock into other molecular structures.

So, when you smell your cup of coffee, the molecules responsible for the delicious aroma have key atomic groups that can interact with the molecular structures in your nose, sending your brain the message that there is coffee nearby.

Before we continue, however, we should note that protein folding is not always perfect, and some proteins may end up a little misshapen. With their particular molecular surfaces, these misshapen

---

[3] If you want to help in solving the protein-folding problem, you can contribute the spare power of your computer when you are asleep. Simply register with folding. stanford.edu and you can rest easy knowing that your computer is working on a major mystery of molecular biology.

proteins are generally inactive, but sometimes the new shape finds a new molecular hole that it can fit in, and these proteins can be active in a brand-new way. However, with this new lease of life, these new proteins are not necessarily beneficial, as some of these are prions, resulting in illnesses such as the brain-wasting Creutzfeldt–Jakob disease.

The secrets of life were laid bare by molecular biology. Molecular machinery churns and grinds within every one of us. Molecules do it all: they are the cogs, the springs, the fuel, the supply truck, the defences, the code, and the decoder. Together, they make cells, join cells into tissues, tissues into organs, and organs into organisms.[4]

THE MACHINERY OF ATOMS

Your molecular machinery is governed by the push and pull of the electromagnetic force on individual atoms. So, how complicated are atoms?

The idea that all material objects are built out of a limited set of building blocks, the atoms, is an ancient philosophical idea, but the realization that the world is actually built this way dawned relatively slowly. The great scientist Ludwig Boltzmann was driven to despair, a despair that contributed to his eventual suicide in 1906, by the continued resistance to his ideas that the properties of gases could be explained by assuming they consisted of a multitude of colliding atoms. Richard Feynman began his legendary *Lectures in Physics* by choosing the most important sentence to be passed on to future physicists: 'all things are made of atoms – little particles that move around

---

[4] When you look at a bowl of petunias, you may marvel at the seemingly immense gap between animal and plant life. But if you want to be brought a little more down to Earth, have a look at the similarities between hemoglobin, the molecule in your blood responsible for carrying oxygen to your tissues, and chlorophyll, the molecule responsible for capturing carbon dioxide in plants. Hemoglobin is a ring-like molecule built around an iron atom, while chlorophyll is a similar ring-like molecule built around a magnesium atom. Look it up, stare at it, and ponder what this is telling us about life on Earth.

in perpetual motion, attracting each other when they are a little distance apart, but repelling upon being squeezed into one another.'

So, taking apart the many molecules that make up a human being, simplicity begins to emerge. Molecules are legion, but built from a limited number of distinct building blocks. In fact, nature has furnished this Universe with only 92 of these fundamental *elements*.[5]

What is an atom? At school, children get a general picture of an atom as a positively charged ball of mass, known as a nucleus, surrounded by a buzzing cloud of negatively charged electrons, making the overall atom electrically neutral. The nucleus is composed of individual bits called *nucleons*: the *protons*, which carry all of the positive charge in the nucleus, and the slightly heavier *neutrons*, which carry no net charge.

A diagram of the atom is shown in Figure 8, which is not to scale. Electrons orbit at distances much, much greater than the size of the

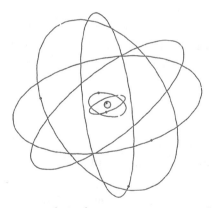

FIGURE 8 A schematic representation of a carbon atom with six electrons 'orbiting' the nucleus. Note that this picture is not to scale. The chemical properties of carbon are set by the fact that there are four electrons in the outermost orbit. It is this part of the carbon atom that interacts with the outside world.

---

[5] While nature has provided 92 naturally occurring elements, other, more massive atoms can be created in the laboratory. However, these have been found to be unstable, rapidly decaying back into the elements of the Universe.

nucleus. This means that most of the atom, and hence all things that are constructed of atoms, including us, is mostly completely empty space. If we think of an atomic nucleus as being about the same size as a fly, the surrounding electrons would be orbiting more than a hundred metres away.[6]

The *only* difference between two elements is the charge of the nucleus, which is determined by the number of protons it contains. Most of the atoms of your acquaintance are electrically neutral, and so have the same number of electrons in orbit about the nucleus. For a particular element, individual atoms may have a differing number of neutrons, giving us differing *isotopes*. These do have an important role in our story, but as isotopes of an element are (roughly) chemically equivalent, we won't worry about the distinction right now.

Here's the big picture. The complexity of the chemical processes within a cell is overwhelming. However, looking deeper, we find that life's molecules are built of a limited number of atomic elements, just 92 of them. And these 92 blocks are in turn built from just three pieces: protons, neutrons and electrons. Like a child's building blocks, a limited number of pieces can be arranged into a multitude of combinations.

THE FUNDAMENTAL MACHINERY

We can go deeper still. The *Standard Model of particle physics* lists the particles that we believe are fundamental, out of which everything else is made. Let's meet the whole gang.

As hard as we have tried, we have not succeeded in splitting an electron into smaller pieces. The electron appears to be fundamental, an ultimate building block of the Universe. However, with powerful atom-smashers (which do exactly what it says on the tin, and are more commonly known as particle accelerators) we have been able to poke the innards of protons and neutrons.

---

[6] For a more detailed and historical look at the unveiling of the structure of atoms, we recommend the rather wonderful book *The Fly in the Cathedral* (Cathcart, 2004).

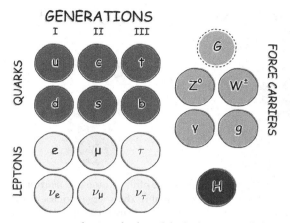

FIGURE 9 In the Standard Model, the leptons and the quarks are the building blocks of all matter, although virtually everything we see in the Universe is composed of up and down quarks, and electrons. These are coupled with the force carriers and the Higgs boson, which is responsible for mass.

Inside, we have found pieces known as *quarks*. As with the electron, we have not been able to smash quarks into smaller pieces, and we think that they too are fundamental building blocks of matter.

Figure 9 shows all the pieces. Inside the proton and neutron we have found the up (u) and the down (d) quark. The electron (e) can be found on the left. The electron has –1 units of electric charge, the up quark has +⅔ units, and the down quark –⅓ units.[7] Together, these three particles make up all the matter with which you are familiar.

To make a proton, you combine two up quarks with a down quark, to give a total charge of +1, and to make a neutron, you add one up quark to two down quarks, with a resulting charge of zero

---

[7] Students can be a little unnerved by the fractional charges of quarks, but it is just a quirk of the fact that electrons were discovered first. It would make more sense to assign the down quark a charge of –1, the up quark +2 and the electron –3, and so the discovery of the electron first has led to significant confusion to physics students. If you don't believe us, ask them which way a current flows in a wire, compared to the motion of electrons.

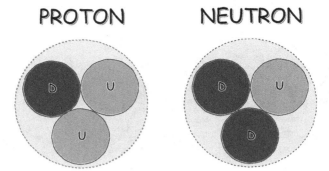

FIGURE 10 The quark constituents of the proton and neutron. While there are six different quarks available, these key particles, which are the bits of the nucleus of every atom in the Universe, are built from only the two lightest quarks, the up and the down.

(Figure 10). The proton and neutron are examples of particles known as *baryons*, made from three quarks bound together.

Once you have protons and neutrons, all you need is to join them to get atomic nuclei, add electrons to get atoms, and onwards to molecules, cells, and human beings.[8]

The astute reader may wonder, 'there are a couple of ways we can combine the up quark and the down quark, so could we join three ups together, or three downs?'

Congratulations, you are now a (novice) particle physicist. Yes, these additional combinations exist. Combining three up quarks gives the attractively named Delta-plus-plus particle (symbolically, Δ^{++}) that has a positive charge of twice that of the proton. The other, the mixture of three down quarks, is the Delta-minus particle (Δ^-), a negatively charged particle with the same charge as the electron. Why don't we see these other combinations of quarks in everyday life? Because these combinations are unstable and rapidly (and we mean rapidly, in about

---

[8] Disclaimer: building a human being from the quarks up is more difficult than this paragraph suggests. We recommend the traditional method.

0.000000000000001 seconds) decay into the normal matter we see around us.

So what of the other particles in the Standard Model? Where did we find those?

A little over half a century ago, particle physics was a mess. While perhaps not as messy as cell biology, new experiments were finding huge numbers of different particles. The Standard Model is the final product of decades of intense labour from physicists, searching for underlying simplicity.

They discovered that more than the up quark and down quark is required to explain all of the properties of the particles we see in particle accelerators. Specifically, we need four more quarks, two more with a charge of +⅔, the charm quark and the top quark, as well as two more with a charge of -⅓, the strange quark and the bottom quark. The up and the down quark are the lightest of the quarks, and as we move along the quarks get more and more massive. These three layers are referred to as the quark generations.[9] Why there are three generations when the Universe seems to only use one is an as-yet unexplained mystery of the cosmos.

Why the goofy quark names? Because we had to call them something. It's not easy naming something you cannot see, and which in our theories is basically a point with a few numbers attached.

Not to be left out, the electron has heavier siblings, namely the muon (μ) and tauon (τ), which make up what are known as the lepton generations. The fact that there are three generations of both the

---

[9] As newly minted particle physicists, your first task is to confirm that there are 56 different ways of grouping three quarks out of a possible six, but the number of particles seen in experiments is many times greater. One reason is that we ignored the mesons, a quark/antiquark pair. Another reason is that protons are, for want of a better word, excitable. Picture the quarks inside a proton orbiting each other. If we can inject some energy, we can change the configuration. This new state would look a lot like a proton – same quarks, same charge – but because it has more energy, it has more mass. Before we realized that these particles were simply excited states, the proton's 20% heavier form was called Δ⁺ ('delta-plus').

leptons and the quarks is probably telling us something pretty deep about the Universe. We just don't know what.

To round off this picture, each member of the lepton family is accompanied by a ghostly, almost massless counterpart, known as the *neutrino* (ν). While produced in copious amounts in nuclear reactions, these 'little neutral ones' (from the Italian) rarely interact and thus stream through matter as if it is not there. In fact, in every second, trillions of neutrinos produced in the heart of the Sun pass harmlessly through your body.

Each particle also has a twin: its *antiparticle*. For every particle that we have observed in a particle accelerator, we have also observed a particle with exactly the same mass but opposite charge. For historical reasons, the electron's antiparticle is called the *positron*; the others just have 'anti' added to the front of their names. Some particles, like the photon (particle of light), are *their own* antiparticles. When a particle meets its antiparticle, they can annihilate into two photons.

In the next chapter we'll look at how forces work, but in terms of building matter, that's it. Those twelve little pieces (and their antiparticle twins) are the ultimate building blocks in the Universe.

We have arrived at simplicity. All of the complexity we see is built out of a handful of fundamental particles, arranged, like Lego® blocks, in a myriad of ways, held together with an even smaller number of fundamental forces. The whole Universe (and any other life form out there) is built the same way.

In the coming chapters, we will consider what happens when we play around with the forces, but we can start by asking how different the Universe would be if we change the masses of the fundamental particles.

CRACKING AT THE FOUNDATIONS

A Lego® set with just three types of brick would demand a lot of imagination from a child, if they were to make something interesting.

And yet, look around: three particles – the electron, up quark and down quark – make up everything we see.

Now that we've described the limited fundamental set of building blocks that create the wealth of structure we see around us – atoms, molecules, trees, people, stars, planets and more – it's time to ask how different the Universe would be if the building blocks were different.

These particles are so simple that they can be described with just a handful of properties, such as mass and charge. The charges, as we saw above, seem to be neat numbers: in units of the electron charge, the quark charges are +2/3 and –1/3. There are other numbers associated with these particles, known as *quantum numbers*, named *spin*, *isospin*, *weak-isospin* etc. These, too, are expressible in nice, neat, whole number chunks. On the other hand, one property of these particles that is not so tidy is the one that we might consider the most important: the particles' masses.

For example, the up and down quarks are 4.5 and 9.4 times heavier than the electron.[10] These aren't nice, neat numbers. And yet, they are fundamental to the Standard Model of particle physics. Frustratingly, we can measure them, but we can't explain them in terms of anything else.

The other paraphernalia of the Standard Model aren't any better. The other four quarks are 190, 2495, 8180 and 338,960 times heavier than the electron, while the muon and tau are 206.768 284 and 3477.15 times heavier.

Is there anything special about the particular values they have? What happens in a universe in which the electron and quark masses are slightly different?

One might think that, since life is so hardy and robust, you'd just get a different form of life. It might not look like us, but since life in our Universe can make use of the hodge-podge of chemical reactions on offer, any old universe would do something. Right?

---

[10] We take the electron, up quark and down quark masses to be 0.510 998 928 MeV, 2.3 MeV, 4.8 MeV, respectively (Olive, 2014).

MASSES OF THE FUNDAMENTAL ZOO

In fact, it is rather easy to arrange for a universe to have no chemistry at all. Grab a hold of the particle mass dials and let's create a few universes.

For simplicity, we will change the mass of only the up and down quarks, the quarks that are the basic constituents of protons and neutrons. You might think that this would simply make heavier protons and neutrons, and hence slightly heavier things in general. However, the picture is a little more complicated than that.

Remember why, despite there being so many types of quarks and so many ways of putting them together, we only see matter built from protons (up-up-down) and neutrons (up-down-down). When we use particle accelerators to make heavier particles, such as the Δ^{++} (Delta-plus-plus, up-up-up), Σ^+ (Sigma-plus, up-up-strange) or even the muon, they *decay* into lighter particles.

Why can heavier particles decay into lighter particles? The answer is in Einstein's famous $E = mc^2$. By multiplying the particle's mass (m) by the speed of light (c) squared, we can calculate how much energy (E) can be released. That energy can be used to purchase new particles, so long as the old particle can afford it.

Consider the example in Figure 11. The Δ^{++} particle has a mass of 1232 in suitable units (these are megaelectronvolts, or MeVs, the particle physicist's energy unit of choice), while the proton and the pion (a *meson*, that is, a combination of a quark and an antiquark) have energies associated with their masses of 938 and 140 MeV, respectively. The Δ^{++} particle has more than enough energy to settle the tab, so the decay into a proton and a pion is allowed. The leftover energy is given to the proton and pion as a little extra *kinetic* energy, the energy of motion. They fly quickly from the site of the transaction.

We'll come to conservation laws in more detail when we take a deeper look at symmetries in the Universe, but having enough energy is not enough to ensure that the transaction goes through. The Δ^{++} decay also conserves what is known as *baryon number*: the total number of quarks minus the number of antiquarks is the same

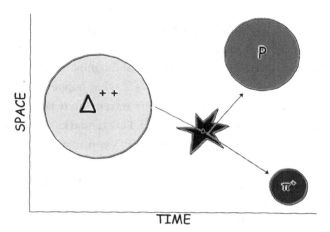

FIGURE 11 The Δ^{++} is more massive than the sum of the masses of the proton and the pion (π^+). This means that it is energetically possible for the Δ^{++} to decay into a proton and a pion.

before and after the decay. The meson, being a quark/antiquark pair, has a baryon number of zero and so does not contribute to the tally. This conservation of baryon number means that, even if it had enough mass, and hence energy, the Δ^{++} could not decay into two protons; baryon number would not be conserved.

So the proton holds a very important title: it is the lightest particle made from three quarks. The proton is stable because there is no lighter baryon to decay into.[11] The buck stops here. The neutron, by contrast, decays. If left on its own, it has about 15 minutes before it becomes a proton, an electron and an antineutrino.

In the early Universe, where there was nothing but a hot, dense soup of particles and radiation, violent collisions were continuously producing and destroying particles of all masses and types. As the Universe cooled, the more massive baryons decayed into protons and

---

[11] There is a subtlety here. If baryon number is not perfectly conserved, then protons will eventually decay. This has never been observed, and the calculated timescale for such decays is many, many orders of magnitude longer than the current age of the Universe. Here, we treat protons as stable.

neutrons. The Universe managed to lock some neutrons away inside nuclei in the first few minutes, before they decayed.

By messing with the up and down quark masses, we can change a lot of this story. We can strip the proton of its 'most stable' title, and even affect neutrons inside nuclei. Let's see what happens.

The Delta-Plus-Plus Universe: Let's start by increasing the mass of the down quark by a factor of about 70. Down quarks would readily transform into up quarks (and other stuff), even inside protons and neutrons. Thus, they would rapidly decay into the new 'most stable' title-holder, our old friend the Δ^{++} particle. We would find ourselves in the 'Delta-plus-plus universe'.

As we've seen, the Δ^{++} particle is a baryon containing three up quarks. Unlike the proton and neutron, however, the extra charge, and hence electromagnetic repulsion, on the Δ^{++} particles makes them much harder to bind together. Individual Δ^{++} particles can capture two electrons to make a helium-like element. And this will be the only element in the universe. Farewell, periodic table! The online PubChem database in our Universe lists 60,770,909 chemical compounds (and counting[12]); in the Δ^{++} universe it would list just *one*. And, being like helium, it would undergo zero chemical reactions.

The Delta-Minus Universe: Beginning with our Universe again, let's instead increase the mass of the up quark by a factor of 130. Again, the proton and neutron will be replaced by one kind of stable particle made of three down quarks, known as the Δ^-. Within this Δ^- universe, with no neutrons to help dilute the repulsive force of their negative charge, there again will be just one type of atom, and, in a dramatic improvement on the Δ^{++} universe, one chemical reaction! Two Δ^- particles can form a molecule, assuming that we replace all electrons with their positively charged alter-ego, the positron.

The Hydrogen Universe: To create a hydrogen-only universe, we increase the mass of the down quark by at least a factor of 3. Here, no

---

[12] pubchem.ncbi.nlm.nih.gov

neutron is safe. Even inside nuclei, neutrons decay. Once again, kiss your chemistry textbook goodbye, as we'd be left with one type of atom and one chemical reaction.

The Neutron Universe: If you think the hydrogen universe is rather featureless, let's instead increase the mass of the up quark by a factor of 6. The result is that the proton falls apart. In a reversal of what we see in our Universe, the proton, including protons buried in the apparent safety of the atomic nucleus, decay into neutrons, positrons and neutrinos. This is by far the worst universe we've so far encountered: no atoms, no chemical reactions. Just endless, featureless space filled with inert, boring neutrons.

There is more than one way to create a neutron universe. Decrease the mass of the down quark by just 8 per cent and protons in atoms will capture the electrons in orbit around them, forming neutrons. Atoms would dissolve into clouds of featureless, chemical-free neutrons.

What about the other particle of everyday stuff, the electron? Since the electron (and its antiparticle, the positron) is involved in the decay of neutron and proton, it too can sterilize a universe. For example, increase its mass by a factor of 2.5, and we're in the neutron universe again.

The situation is summarized in Figure 12. Each point in the plot represents a different universe; throw a dart and you've chosen the mass of the electron and the difference between the down and up quark masses. The grey regions show where the complex chemistry of our Universe would simply not occur, with universes dominated by only protons (the hydrogen universe), or only neutrons. On the left there is a white wedge of acceptable universes, universes in which chemistry would offer the basic building blocks of life, while the black dot inside marks the location of our Universe. As you can see, we luckily find ourselves in this relatively small chunk of the picture, in a place where we can exist and question the nature of the Universe. This, effectively, is the start of the fine-tuning journey.

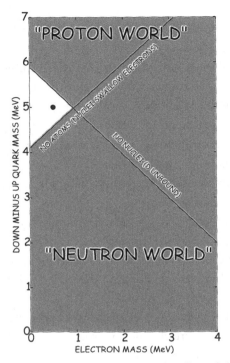

FIGURE 12 This summarizes the effect of changing the mass of the electron and the difference in the down and up quark masses. The region is sliced up into universes with uninteresting chemistry, shaded in grey, while the small white region allows chemistry for life. The black dot is our Universe. (Based on a figure from Hogan (2009).)

An important part of this plot to keep a firm eye on is the limits of the axes. We've plotted the quark difference from 0 to 7 MeV, and the electron mass from 0 to 4 MeV. This region is the interesting bit of the plot; the rest is boring and grey. If we wanted, instead, to get an impression of the *smallness* of the white, life-permitting region, we could extend the axes up to the mass of the largest quark we've observed: the top quark. To see this, add 4 kilometres of grey paper to the right, top and bottom of the plot. (If you make this 10,000 acre plot, please send us a photo.)

If we extend to the limits of our experimental data, we'd mark 6.5 million MeV at the ends a plot that could cover most of Tasmania, or all of Switzerland, or half of South Korea, or most of West Virginia, or two Death Stars,[13] depending on which continent/universe you inhabit.

The maximum mass that our theories can handle is called the *Planck mass*. At this mass, a particle is predicted to become its own black hole. We don't currently have a theory of quantum gravity – that is, a theory that knows what to do with quantum things (like particles) in the grip of their own gravity. So the Planck mass is a firm limit to our physical theories. And if you can imagine extending the plot above by a few tens of thousands of light years on each side, then you're doing better than us.

As you can imagine, this is the first step in a long journey, with much more to come. And we have not finished with the masses of the fundamental particles. Stars, the source of the energy that powers our existence, are driven by nuclear reactions. As we'll see in the next chapter, messing with the properties of quarks and electrons can dramatically affect the stability and lifetime of the energy source of life as we know it.

GIVING YOUR ATOMS PERSONALITY

As well as mass, particles such as electrons and quarks carry other properties that dictate their interactions with other particles. One of the most important, and strangest, is particle *spin*.

As its name suggests, a particle's spin is like the spin of the Earth, or like a child's spinning top.[14] There are several key differences, however, which significantly influence the way the matter in the Universe behaves.

---

[13] That's two original (*Star Wars: A New Hope*) Death Stars. The size of the Death Star II (*Return of the Jedi*) is the subject of passionate Internet debate.

[14] Do children have spinning tops anymore? Most children these days appear to have their faces glued to iPads and iPhones. It wasn't like that when we were young (more's the pity!).

Firstly, spin is *quantized*, which means a particle may only have a whole number of indivisible chunks of spin. While some particles lack spin, and so are said to be spin zero particles, all electrons have exactly the same amount of spin, namely one-half (again, in appropriate units). Other particles have one unit of spin, others again with spins of three-halves, with increments of half a unit of spin.

This might sound like mere bookkeeping; what does it have to do with the workings of the Universe? Well, physicists divide particles into two families: those with an integer value of spin (0, 1, 2, . . .) which are called *bosons*, and those with half-integer spins (½, 1½, 2½, . . .) which are known as the *fermions*.[15] Of course, the most famous boson in the world is the Higgs boson, but there are others, as we will see shortly. Why the two families?

It's to do with the way particles pack together. Bosons are flexible little things: in a small box, we can continually pack in more and more bosons, and there is nothing to stop us. The most familiar boson in our everyday life is the photon, the particle of light, which has a spin of 1. If you're trying to pack a box with photons, each new addition doesn't much mind if there are already photons inside.

Fermions, however, do not play by the same rules. Fermions don't like other similar fermions, and once they have established *their* space, they don't let anyone else join them.

Anyone who has tried to lay their towel on a crowded beach will have encountered this, as straying too close to established beach-goers produces awkward stares. And if the beach is full, then too bad.[16] So it is with fermions. If you try and pack fermions into a box, eventually the box will be full and you will be unable to squeeze any more in.

---

[15] Bosons are named after Indian physicist Satyendra Nath Bose. Physicists pronounce the word 'bow (as in bow-and-arrow) – zon', although the media sometimes seems to think they have some sort of nautical origin, referring to them as bow-suns (we're looking at you here, ABC News24 Breakfast Team). The media never seem to mention the boson's counterpart: the *fermion*, named after Italian physicist Enrico Fermi.

[16] Of course, we could have used many examples from nature to illustrate this point, such as a seabird on a crowded cliff-face, or lions battling it out for mating rights with a pride, but it is useful to remember that we humans are not that distant from our animal cousins!

The most common fermion to our everyday life is the electron, and this packing limitation of electrons has a profound influence on our world, effectively defining the subject of chemistry. This is the famous *Pauli exclusion principle* in action.

Remembering your atomic physics, electrons live in certain energy levels about atoms, with two in the lowest level of an atom, known as the ground state. Why two? Electrons possess a quantized spin of ½, which can point either upwards or downwards, and so the lowest energy state can be occupied by one electron with its spin pointing upwards, as well as a second electron with its spin pointing downwards. (Classically, we define the *direction* of a spin with the right-hand-rule: wrap your right hand around the spinning object, with your fingers pointing with the spin. Give a thumbs-up; this defines the direction of the spin). In terms of packing electrons, this lowest level of the atom is full, and any additional electrons must go into higher levels. And as these upper levels fill, incoming electrons settle into higher and higher energy levels.

What does this have to do with chemistry? Well, chemistry is about how atoms interact, and this is completely defined by the locations of the electrons in the outermost energy levels; these are the most loosely bound electrons, and it is these that can be swapped between atoms to allow them to bond and form complex molecules. It is these electrons, and the properties of their spin and orbits, that give atoms their personality.

The same is true of atomic nuclei. Protons and neutrons, themselves fermions, can be found only on distinct energy levels within the nucleus, packed in accordance with the Pauli exclusion principle. This explains why, while isolated neutrons rapidly decay, a neutron *inside* an atomic nucleus is effectively stable. It cannot decay because no lower energy positions are available for the resultant proton.

What if electrons were bosons rather than fermions? The result would be disastrous as there would be nothing to prevent all the electrons occupying the lowest level in an atom, like stuffing as many photons in the box as you can. Once again, wave chemistry – and the chemical complexity and flexibility needed by life – goodbye.

These bosonic electrons would be very tightly bound to their nuclei, with little inclination to be shared with other atoms. This universe would be a sea of individual atoms floating through the cosmos, minding their own business and not getting involved in this messy work of forming molecules.

However, the situation could be even more complicated! As well as the electrons, the quarks are also fermions, also possessing a spin of one-half. And this spin is imprinted (in a rather complex fashion) onto the particle they form. And both the proton and the neutron, even though they are composite objects, have a total spin of one-half, also making them fermions, and ensuring they obey the fermion rules of packing.

This means that the protons and neutrons of an atom are arranged in orbits very similar to the orbits of the much more distant electrons. If quarks, and so also the protons and neutrons, had an integer spin, just like the electrons, there would be nothing to prevent all of them collapsing down and occupying the lowest energy level. Collapsing all of the protons and neutrons in your atoms is not necessarily a catastrophe; there are more important things to worry about.

MELTING THE SOLID STATE

There is another way that electrons can bring about detrimental changes to our Universe. When two atoms or molecules come together and shuffle electrons, a chemical reaction has occurred. Sometimes, the electron is not pilfered but shared: as the two atoms or molecules wrestle for possession of the electron, a chemical bond is formed.

In solids, atoms are held together by chemical bonds in a fixed lattice. Imagine a three-dimensional grid of balls held together by springs, something like the one in Figure 13.

We can break the lattice by shaking it vigorously. Similarly, we can melt a solid by heating it up. Heat consists of microscopic

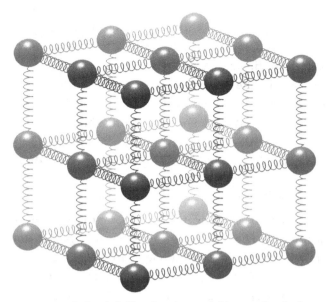

FIGURE 13 A simple ball-and-spring model for a molecular lattice. Atoms are held in a repeating crystal structure, and vibrate as the bonds between them stretch and compress. Heating the solid causes the atoms to vibrate faster and further, straining the bonds. If enough heat is added, the bonds break and the substance melts. The melting point of the lattice depends upon the masses of the fundamental particles and the strength of the fundamental forces. Changing these can make the springs break more easily, or the atoms vibrate more violently, melting solids into liquids.

random motions, and too much jiggling of the lattice will break the chemical bonds between the atoms. The atoms move too quickly for any particular bond to hold for very long, and the material loses its rigidity, turning into a liquid or a gas.

There is a natural, unavoidable jiggling in all objects, which comes from quantum mechanics. Unlike heating, this jiggling is a fundamental property of the quantum world, preventing us from ever cooling something to absolute zero as this would require the atoms in our lattice to be completely still. In our Universe, this jiggling is relatively gentle and won't melt objects so long as electrons are much lighter than protons. The quantum jiggling will cause electrons to buzz around the nuclei, but they're too light (1836.15 times lighter than the proton) to knock nuclei out of the lattice. And so, solid materials stay solid as long as the temperature is sufficiently low.

However, if the electron mass were within a factor of a hundred of the proton mass, the quantum jiggling of electrons would destroy the lattice. In short: no solids. No solid planets, no stable DNA molecules, no bones, no semi-permeable living cell walls, no organs. A complete mess, and almost certainly no life!

THE TROUBLE WITH HIGGS

You'd have to have been living under a rock for the last few years not to have heard of the Higgs boson. If you are not a physicist, you might have heard talk of the long-sought 'God particle',[17] which somehow gives mass to fundamental particles. *Time* magazine even nominated the Higgs boson as a possible 'Person of the Year'. Unfortunately, each of the five sentences of the nomination contained at least one serious physics error.[18] We'll try to clear things up.

The discovery of the Higgs particle, while an immense success for particle physics, brings its own fine-tuning headaches. The problem is not with what has been discovered in particle accelerators, namely the Large Hadron Collider, but with what hasn't been seen! This is going to take a bit of explaining, so let's take a step back.

We've already been introduced to the Standard Model of particle physics, describing the building blocks of matter and radiation in the Universe. The mathematical language in which the Standard Model is written goes by the scientifically sexy name of *quantum field theory*. Discovering the right equations occupied the careers of a heroic

---

[17]   The media continually refer to the Higgs boson as the 'God particle' after this name was used in the title of a 1993 popular-level book by Leon Lederman and Dick Teresi. The Higgs boson was originally called the 'God-damn' particle, because it was so hard to find. No self-respecting physicist would refer to the Higgs as the 'God particle' if there was another physicist in earshot, and we cringe when a newsreader does so.

[18]   Search blogs.scientificamerican.com for *Nomination for Higgs Boson Riddled with Errors*.

generation of physicists. The result of these labours is nothing short of remarkable. Quantum field theory can boast of predictions that are accurate to better than one part in a billion.

In the 1960s and '70s, however, the picture still wasn't complete. Particle physicists were exploring ideas that they hoped would unify the forces of physics, showing that electromagnetism and the weak nuclear force – of which we will learn more in the next chapter – are in fact different manifestations of one fundamental force. The equations, however, had a few undesirable consequences. A new particle – massless, spinless and electrically charged – seemed to be required, but is not observed. In fact, several of the key features of particle physics, namely the gauge bosons, would have to be massless.

These problems were solved by what is known as the *Higgs mechanism*, after the English physicist Peter Higgs. As usual, the discovery was not made in a vacuum; a number of physicists provided clues to and glimpses of the final solution. Higgs himself refers to the 'ABEGHHK'tH mechanism', after the physicists Anderson, Brout, Englert, Guralnik, Hagen, Higgs, Kibble and 't Hooft.[19]

Part of this solution is the postulation of a new *field*. Fields are central to modern physics. By definition, they attach some physical quantity (or set of quantities) to every point in space and time. You can think of the temperature in a room as an example of a *scalar field*, a field in which every point is associated with an individual value. More complicated *vector* fields describe magnetic and electric fields, attaching a value and direction to every point in space and time. Yet more complex physical phenomena are described by more complex forms of fields.

The particles of the Standard Model of particle physics get their mass by interacting with the *Higgs field*. In particular, fundamental

---

[19] Frank Close's historical account in *The Infinity Puzzle* (2011) is highly recommended for its lucid explanation of the physics and skilful untangling of who discovered what.

particles get their *inertial mass* from the Higgs field. Think of an elephant on roller-skates: inertia is what makes it hard to push when it is stopped, and hard to stop when it is moving. You can think of the Higgs field as filling space with syrup. Particles that interact with the field are slowed down, thus behaving as if they have mass.

Note that the Higgs mechanism gives mass to the fundamental particles only. In composite particles, such as protons and neutrons, the individual quark masses make up only a tiny fraction of the mass. The remainder is in the form of the energy that binds the quarks together.

So, the Higgs field is responsible for the mass of fundamental particles. Moreover, the field itself can vibrate, and these waves behave like particles. The particle in question is the *Higgs boson*, whose discovery was one of the key scientific achievements in recent years. The Higgs boson is about 133 times heavier than a proton, with a mass of ~125 GeV (again, in particle physics units) making it a quite massive member of the particle family.

Success all around, with scientists dancing in the street!

Not so fast!

Here's where the fine-tuning headache begins. In our quantum mechanical view of the Universe, empty space is not truly empty, but seethes with quantum fluctuations, with particles popping in and out of existence. Yes, it sounds like something dreamt up in an opium haze, but we need to include this fluctuation of energy on small scales to accurately account for our observations of the Universe.

So, when we talk about the mass of a particle, there are two contributions. Firstly, there is the intrinsic or bare mass. Secondly, there is the constant buzzing of these quantum hangers-on. Each particle is hauling not just itself, but a cloud of vacuum fluctuations. When we measure the mass of a particle, we get both contributions.

For the electron, though there are an infinite number of extra contributions, when summed up, they make only a small difference to the total mass.

But for the Higgs, things are not so simple. If we play the same game and add the contributions from the vacuum, they don't get smaller, and when you add them all up, the mass of the Higgs boson we should measure would be infinite. Clearly something is wrong.

When faced with such *divergences* to infinity, physicists look for a place where we can stop adding all of these contributions. There is a hard upper limit, the *Planck energy*. We don't have a quantum theory of gravity, so extrapolating past the Planck energy (or equivalently, Planck mass) is pointless. Adding this cut-off drastically reduces the expected mass of the Higgs boson from infinity down to about 10^{18} GeV! Getting closer, but still a long, long way from the observed value.

Since the masses of the fundamental particles of the Standard Model are tied to the Higgs field, if the Higgs mass were approximately equal to the Planck mass, then all of the Standard Model particles would be similarly Planck–proportioned.[20] And, as we have seen, increasing the fundamental particle masses by even a factor of a few is a disaster for life; increasing by 10^{16} would be . . . well, don't look. It's not pretty.

So, we suspect that we're missing something, a physical effect that wipes out the additional mass added from quantum fluctuations, and does so very precisely. Wiping out a factor of 2 is not going to save the day, as the predicted Higgs mass would still be immensely larger than is measured. A factor of a hundred or a thousand does not help. Nor a billion, nor a trillion. No, we need to balance the contributions from the vacuum by a factor of about 10^{16}.

[20] A technicality: the masses of the fundamental particles of the Standard Model are tied to the Higgs 'vacuum expectation value', v. This sets the typical mass scale of these particles; their specific masses are roughly v multiplied by a 'Yukawa parameter', Y. For example, the electron's Yukawa parameter is 2.6×10^{-6}. The up and down quarks' Yukawa parameters are similarly tiny, which underscores the fact that 'it's the up and down quarks that are absurdly light', in the words of Leonard Susskind (2005, p. 176). The Higgs mass is closely related to v via a dimensionless parameter called the *quartic coupling*, which has a value of 0.13. We are assuming that a large value for the Higgs mass would result in a large value of v and hence large particle masses. For this to be false, either the quartic coupling or the Yukawa parameters would have to be finely tuned.

Particle physicists are made of stern stuff, and have searched high and low for potential solutions. Their favoured one is known as *supersymmetry*.

The idea is quite simple. For every member of the particle zoo, we hypothesize that they have supersymmetric cousins, affectionately known as *sparticles*. For each fermion, their supersymmetric cousin is a boson, and vice versa. For fermions, such as the electron and quarks, the naming convention is to prefix an 's', giving selectrons and squarks. For bosons we add '-ino' to their name, so the W and Z bosons have cousins known as Winos and Zinos. This can make presentations about supersymmetry sound very odd.

What is the attraction of supersymmetry? If supersymmetry holds, the vacuum contributions of particles to the Higgs mass are exactly *cancelled* by their sparticle cousins. Fantastic! Supersymmetry appears to save the day! But there is one, tiny drawback: that there is no observational evidence for any of the supersymmetric partners. None. Zero. Zilch.

This means the supersymmetric partners, if they exist, must be very massive, more massive than the energy available at the Large Hadron Collider, our biggest particle accelerator. If we can concentrate enough energy in one place, then Einstein's $E = mc^2$ implies that we should be able to create particles with the equivalent mass. This is why we have had to search long and hard for the Higgs boson: its large mass required the power of the souped-up Large Hadron Collider.

Hence, only when there is enough energy around to make these massive supersymmetric partners does the Universe actually behave in a supersymmetric fashion. So, supersymmetry only cancels out the contributions to the Higgs mass that are larger than this *supersymmetry mass (or energy) scale*. The large mass difference between the quarks and the squarks, and the leptons and the sleptons, makes it hard to envisage that the mass contributions from the various quantum fluctuations will neatly cancel. The fact that the Higgs boson is comparatively so light currently haunts modern particle physics. Some worry that supersymmetry is dead and that we need to go back to the drawing board.

So, the mass of the Higgs boson presents us with a conundrum. Our quantum mechanical calculations predict that it should have a mass of 10^{18} GeV. Life requires a value not too much different to what we observe, so that the masses of the fundamental particles are not disastrously large. There must be an as yet unknown mechanism that slices off the contributions from the quantum vacuum, reducing it down to the observed value. This slicing has to be done precisely, not too much and not so little as to destabilize the rest of particle physics. This is a cut as fine as one part in 10^{16}. Maybe there is a natural solution to this cutting, but it seems quite lucky for our Universe that the slicing resulted in stable particle physics. This problem – known as the *hierarchy problem* – keeps particle physicists awake at night.

THE TAIL IS WAGGING THE DOG

We've looked at some of the basic properties of our Universe – the masses of the fundamental particles that make up matter, and the spin of the electron and quarks. We have found that if these properties had been slightly different, the intricate physics and chemistry of our Universe would not exist. These other universes have a vastly diminished ability to form the complex molecules that are essential for life as we know it, or even life as we can imagine it.

In our day-to-day lives, we tend to forget the extremes of small and large upon which we depend. We are small creatures on the thin, solid surface of a still mostly molten planet, and below our feet sits a hot, spinning ball of metal at the Earth's core.[21] Inside, we are made of atoms, trillions of trillions of them in an astonishing concert of stability and motion, structure and energy. An individual atom is unimaginably small, and its protons, neutrons and electrons smaller

---

[21] http://xkcd.com/913/

still. Their properties might seem like mere textbook technicalities, and yet their role in how the Universe plays out is enormous. The smallest ingredients of the Universe dramatically affect its bigger structures, especially the chemistry of life.

It really is a case of the elemental tail wagging the cosmic dog!

3   Can You Feel the Force?

In the preceding chapter, we looked at what a human being is made of. Here, we look at what animates us. In particular, we consider the fundamental forces of nature: the rules that tell our basic parts how to interact with each other, that cause motion and change. Once again, we'll see what happens when we bend the rules.

This chapter and the next will consider two distinct, but ultimately intertwined concepts: forces and energy. To the physics novice, they might seem quite different. We experience forces in the pull of gravity, the thud of a rugby tackle, and snap of a twig underfoot. Energy, on the other hand, flows through wires, is supplied by food and spent on the treadmill.

In classical mechanics (that is, the physics of motion), problems can be approached by calculating forces or by calculating energy. Consider (as usual) a block sliding down a frictionless slope: what is its speed when it hits the bottom? We could keep track of the forces, adding each little push and pull. Or, we could follow the transformation of its stored *potential* energy into motion, noting that the total amount of energy remains the same. Both approaches give the same result, but sometimes one is easier to calculate.

We'll begin by examining the forces of the Universe.

THE FUNDAMENTAL FORCES

The Universe contains a wide variety of pushes and pulls, so it is perhaps surprising that, at bottom, a mere *four* fundamental forces govern the lot. In everyday life, we experience only two of these: gravity and electromagnetism. The other two are called the *strong nuclear force* and the *weak nuclear force*, and unless you care deeply for the very centres of atoms, you won't meet them day-to-day.

In the previous chapter, we met the basic Lego® bricks of the Universe: the quarks and the leptons, the left side of the table of the Standard Model (Figure 9). The other side of the table lists the glue. These are the *gauge bosons*, the carriers of the forces.

The electromagnetic force is 'carried' by the photon, γ. Similarly, gravity is thought to be carried by the as yet undiscovered graviton, G. The strong nuclear force is carried by the massless gluons, so named because they glue the nucleus together. Finally, the weak force is the most enigmatic of the four, carried by three massive particles: Z^0, W^+ and W^-. It is responsible for some of the bizarre properties of this Universe.

Like electromagnetism, the strong force can act over indefinitely large distances. However, the electrical neutrality of your body and that of your friend means that you don't feel a net attraction or repulsion (at least, not due to electromagnetism); you are shielded from the truly immense force of electromagnetism. Similarly, we can think of protons and neutrons as being 'strong neutral', with the true power of the strong force felt within their interiors. However, when protons and neutrons get close enough to feel each other's quarks, the strong force leaks between them, binding the nucleus of an atom. The strong force is effectively a *short-range force*. It is quite startling to think that the stability of your atoms is due to a residue of the true strong force.

Let's take a look at these forces and their particles in detail.

Feynman Forces

How can a particle 'carry' a force? Forces are pushes and pulls, so how do charged particles, such as electrons, 'feel' the electromagnetic force with photons?

Several brilliant scientists revealed the answer in the late 1940s, but the name most often tied to this theoretical insight is Richard Feynman. Feynman was struggling with a theory due to Paul Dirac, known as quantum electrodynamics (QED). QED is ambitious, bringing together quantum theory, relativity, the electromagnetic force and the electron. In QED, both electromagnetism (and its photon) and the electron are represented using fields.

In the 1930s and '40s, QED was famously flawed. Perfectly reasonable questions to ask of the theory, such as the probability of this or that process, were giving infinite answers. QED's initial promise was in serious trouble.

Feynman was particularly bothered by the electron. When introduced to the complex world of electromagnetism, students learn that electrons, as charged particles, are a source of *electric field*. We can imagine an electron as being surrounded by the field it creates. If we place another electron in the field of the first, the presence of the field will induce a force on the new charge. Not to be outdone, the newly introduced charge has its own electric field that creates a force on the original charge. The overall effect is that the charges repel one another.

There is a subtlety that is rarely discussed in introductory courses on electromagnetism. Any charge will feel a combined force from the electric fields of all the charges surrounding it, but will not feel the influence of its own electric field. Adding this self-interaction in the solutions to our equations creates infinities, making the mathematical framework of electromagnetism useless. The same problem reappears in QED, rendering the theory physically meaningless.

What Feynman did next is the kind of thing that great physicists do, the kind of thing that opens the door to new understanding about the Universe. He looked for a totally new way to view the theory, one that was more physical, more intuitive, closer to the experimental data but still rigorously rooted in the equations of QED. Feynman got rid of the classical notion of the field. He instead drew pictures.

Before Feynman, physicists such as Julian Schwinger and Sin-Itiro Tomonaga (with whom Feynman shared the 1965 Nobel Prize in Physics) had attempted to solve the equations of QED using mathematically complicated methods. Calculations could take months, even years. Feynman, on the other hand, was trying a new approach to the equations that looked like cartoons. Figure 14 shows an

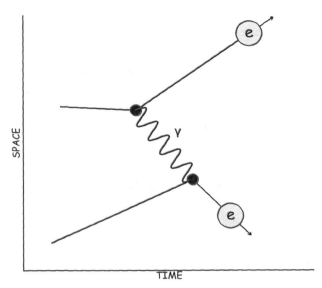

FIGURE 14 Feynman diagram of the electromagnetic force. Unlike classical electromagnetism, where the force is transmitted by an all-pervasive field, in the Feynman picture the force is mediated by the exchange of a force particle, in this case the photon.

example of what is known as a *Feynman diagram*. We can simulate the passage of time by covering the image with a sheet of paper, and then slowly sliding it to the right. The two lines on the left represent electrons, which come together, exchange a photon (squiggly line), and move apart.

You might be wondering: how could a fundamental particle emit, or even be transformed into, a different fundamental particle? Wouldn't that mean that it was made out of smaller bits and pieces? For example, how can an electron emit a photon, unless it already contains a photon? The answer is that the fundamental stuff of particle physics is *quantum fields*, which wobble and wave and fill the Universe. When a quantum field waves in a certain way, it behaves like a particle, or antiparticle, or a collection of particles. There is one special way that the field can be arranged in which there are no particles – the *vacuum*. But the field is still there. When an electron emits a photon, this is a shorthand way of saying that two fields have interacted, spawning a new photon wave.

So beneath Feynman's cartoons was a rigorous mathematical method of dealing with quantum fields, a method that was producing the right answers. Not only that, but Feynman's approach was also much quicker: he once astonished a physicist named Murray Slotnick by, upon hearing of a calculation that had taken two years, redoing it himself overnight. However, Feynman struggled to have his pictures taken seriously – or, indeed, understood at all – until Freeman Dyson proved in 1949 that Feynman, Schwinger and Tomonaga's methods were equivalent. Feynman's ideas are now a staple of modern physics.[1]

So, according to Feynman, how do forces force? Let's go back to Figure 14, where two electrons are approaching each other head on. In the classical picture, the electrons are surrounded by their electromagnetic field, and feel each other through this field. As they get closer, the field (and thus the force) gets stronger and stronger, until the electrons eventually come to a halt and retreat.

In Feynman's picture, things are quite different. As the electrons close in on each other, quantum rules state that there is an increasing chance that one electron will spit out a photon that gets absorbed by the other. When this happens, the electrons change their paths.

As they approach and recede, our electrons share a host of photons. To us, the paths of the particles appear to be smooth, but in detail the particles do a U–turn via a series of sudden shoves.

We can now properly introduce the four fundamental forces via their *Feynman vertices*, where matter lines meet force carriers (Figure 15). We construct the history of particles in the Universe out of these vertices. We've already seen an example of the electromagnetic force in action: a charged particle can emit or absorb a photon. The strong force involves the emission or absorption of a *gluon*; it looks a bit like the electromagnetic vertex, except we draw a spring to

[1] As mentioned in the last chapter, Frank Close's *The Infinity Puzzle* explains all this physics and history beautifully.

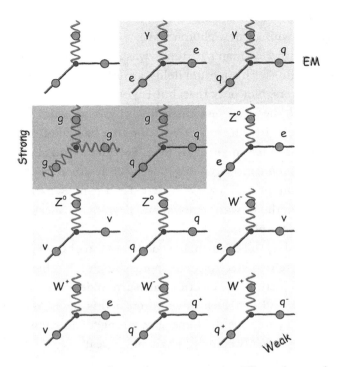

FIGURE 15 A catalogue of Feynman vertices. The twelve panels show some of the ways in which the particles of the Standard Model interact with each other. The top-left image is a basic vertex, with a fermion (solid lines) interacting with a boson (wiggly lines). The electromagnetic interactions (EM) connect charged particles, such as quarks and electrons, interacting via photons (γ). The strong interactions are between quarks and gluons, and even gluons amongst themselves. The Z^0 of the weak interaction interacts with all of the fermions, including the uncharged neutrino (ν), but the $W^\pm$ are much more fun, changing an electron to a neutrino, or changing a quark from having a charge of $+\frac{2}{3}$ to a charge of $-\frac{1}{3}$ (and vice versa).

denote the gluon and it doesn't connect leptons. Just to complicate things, gluons can also interact with gluons.

The Z^0 particle is similar again, pushing and pulling. The W vertex can do more, transforming the particles with which it shares a vertex. An electron, muon or tauon is transformed into a neutrino, or vice versa. Up-type quarks (up, charm and top) are flipped with down-type quarks (down, strange and bottom). And, for another complication, the weak force carriers interact amongst themselves.

Most physicists believe that we can extend this picture to include gravity, with the force of gravity conveyed by the *graviton*. However, refusal of gravity to play ball with the other forces of the Universe is a major stumbling block in modern physics, as we will see.

Feynman Couplings

In the opening chapter of this book, we mentioned the *coupling constants*, the numbers that dictate the strength of a force. In Newton's gravitational theory, the constant G gives the force between a pair of masses a certain distance apart. In Einstein's theory of gravity, G tells us how strongly matter and energy curve spacetime. The other fundamental forces also come with coupling constants.[2] Where do these constants fit into the picture of forces given to us by Feynman?

Feynman's processes are governed by the rules of quantum mechanics, which are probabilistic. The coupling constants tell us the probability that a force particle will be exchanged. Here's an example.

Let's consider the two up quarks inside a proton. Both are positively charged, and so we know from the laws of electromagnetism that they will repel one another. But unlike the electrons we considered in Figure 14, quarks also feel the strong force, which attracts them towards each other (Figure 16). In an instant of time, what is more likely, the exchange of an electromagnetic photon repelling the two quarks, or the exchange of a gluon attracting them?

This roughly depends on the ratio of the coupling constants of the strong and electromagnetic forces. The strong force coupling constant is $\alpha_S \approx 1$, whereas for electromagnetism the coupling constant, also known as the fine structure constant, is $\alpha \approx 1/137$. It is actually

---

[2] Actually, the coupling constants can be thought of as depending upon the energy of the interaction between particles. When we talk about changing the constant, we mean its value at a particular energy.

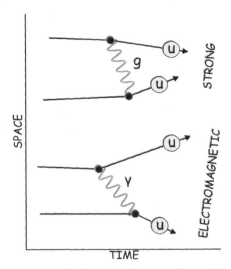

FIGURE 16 The attractive strong force and repulsive electromagnetic force on up quarks inside a proton or neutron, exchanging gluons and photons, respectively. As many more gluons are exchanged per second than photons, the quarks remain stuck together inside the protons and neutrons, overwhelming the electromagnetic force.

the square of the ratio that is important, so an attractive strong force gluon being exchanged is almost 20,000 times more likely than the exchange of a repulsive electromagnetic photon. This means that, when buried inside a proton, the strong force is dominant, whereas the electromagnetic force plays a minor role.

As we've noted, nothing in modern physics tells us why the coupling constants have the values that they do. We have no fundamental theory to predict them, and the best we can do is determine their values experimentally. And we will see that if the Universe had been born with different values of these fundamental constants, things could have been very different indeed!

Chemical Interactions

We've already seen a variety of ways to make chemistry much simpler – too simple for life. We can also make it much more complicated.

The ancient alchemists attempted to transform base metals into precious ones. Even the great scientist Isaac Newton was known to dabble in this strange mix of chemicals, cauldrons and the occult. While the ancients laid the groundwork for modern chemistry by relentless controlled experimentation – producing, along the way, alloys, acids, pigments, and elements from arsenic to phosphorus to zinc – they failed to make gold from lead. We now know why.

Lead and gold are elements. Their differences stem from the different make-up of their atomic nuclei. At the centre of a gold atom is a nucleus containing 79 protons and around 118 neutrons. At the centre of a lead atom, we find 82 protons and around 126 neutrons. So, if we could just reach into that lead nucleus and pull out 3 protons and 8 neutrons, voilà! Gold!

Here's why the alchemists were doomed to failure. When you light a fire, or mix some potions in a cauldron, you are causing chemical reactions. A chemical reaction, by definition, is a rearrangement of the electrons that orbit around nuclei. In terms of mass, an electron trying to upset an atomic nucleus is like a chicken trying to push a hippopotamus.

But suppose the chicken – we mean electron – is moving very fast. Perhaps if we light a super hot fire it will send an electron so fast into the nucleus that it could shift a few of those protons and neutrons. Now a further problem arises. Not only are they heavier, but also the particles in the nucleus are held by the strong nuclear force. The energy needed to break a strong nuclear bond in a nucleus is about 20,000 times higher than an electromagnetic bond. Combined with the fact that the proton is over 1,000 times heavier than the electron, chemical reactions fall millions of times short in energy of disturbing the nucleus. Build all the fires, stir all the cauldrons you like – you won't turn lead into gold.

But what if we changed the rules?

With some tweaking of the free parameters of the Standard Model of particle physics, we could create a universe in which there was no energy gap between *chemical* and *nuclear*. If you left your cake in the oven for too long, you might end up with a burnt cake, or you might end up with a lump of lead.

Our previous meddling with chemistry resulted in a much simpler universe. Erasing the chemical–nuclear energy gap, on the other hand, makes the universe much more complicated. Possibly, the universe would be too chaotic, too unstable for life.

Life needs a balance of stability and motion. We need our hearts to beat – not to sit still (too stable) or pulse erratically (too chaotic). We can't just sit around doing nothing – we'll die, decay, and be eaten by the life forms that don't just sit around. We also need the information in our DNA to be stable, so that new cells follow the same blueprint as our existing cells.

In a universe with no chemical–nuclear energy gap, life could not rely on stable, predictable chemical properties. Oxygen would remain a useful source of chemical energy for life, but if some oxygen atoms collided too heavily with the walls of your lungs, the resultant nuclear reaction could fill your lungs with argon (useless), or arsenic (poisonous).

Every organic reaction network – already astonishingly complicated in our Universe – would be further complicated by scores of available nuclear reactions, each radically altering the chemical properties of the constituents of your body. It would be much more difficult to shield DNA from outside meddling if one could not rely on the chemical properties that make up the strand itself.

And yet, because of this dramatic increase in complexity, it is difficult to predict exactly what *would* happen in such a universe. Perhaps life would find a way, intricately weaving both chemical and nuclear reactions to produce living systems unlike any we can imagine.[3] Regardless, this case illustrates some of the dangers of too

---

[3] Note, of course, that evolution won't help. It needs a self-replicating life form to start with. If primitive cells all dissolve in a flurry of nuclear reactions, natural selection won't have anything to select.

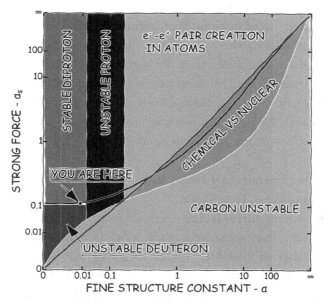

FIGURE 17 The fine-tuning of the electromagnetic and strong forces. Playing these forces off against one-another has a drastic effect on the universe, with an almost imperceptible region of stability near the black dot that denotes our Universe. (Based on Tegmark (1998).)

much complexity for living systems, which rely on the faithful storage and replication of information (Figure 17).

Big Bang Nucleosynthesis

Given the importance of chemistry,[4] we must ask: where do the differing elements around us come from? Over the last one hundred years or so, we've discovered that the Universe has two elemental forges. The first is the Big Bang, which forged the light elements in the first few minutes of the Universe. The second is stars, nuclear furnaces held together by their own gravity. For both, the resultant mix of elements reflects a battle between all four forces of nature. Let's start by looking at the Universe.

[4] As physicists, this was difficult to write.

We will discuss the Big Bang theory in much more detail in Chapter 5; here we focus on its nuclear furnace. As a physical theory, the Big Bang tells us the condition of the Universe between a few seconds and a few minutes after the beginning. In this brief period, the temperature in the Universe was just right to build nuclei from protons and neutrons. This cosmic oven cooked up 75 per cent hydrogen (by mass), 25 per cent helium, and sprinkles of deuterium, lithium and beryllium, largely in agreement with the Universe we see around us.[5]

All four fundamental forces have a say in how much of each element is produced. Here's how.

The strong force is the nuclear glue, so it is obviously involved. Electromagnetism is part of what makes joining two nuclei together so difficult. Nuclei are positively charged and so repel each other. If they are moving too slowly, they'll be pushed apart before the (short-range) strong force has a chance to stick. We need nuclei to be moving very quickly to get close enough for the strong force to reach out and take hold. In other words, we need heat. But not too much heat. The strong force isn't infinitely strong, so a violent enough collision will tear apart the nucleus.

The weak force, too, is crucial. We cannot make nuclei from protons alone; we need neutrons. However, neutrons are unstable and will decay within about 15 minutes, so we need a fresh supply made and bound into nuclei within that timeframe. Thanks to the weak force, and its unique ability to turn up quarks into down quarks and vice versa, so long as the Universe is hot and dense enough, reactions such as 'proton + electron → neutron + neutrino' will keep the neutron stocks up. Nuclear reactions can then proceed in an orderly fashion, building deuterium (proton + neutron), and then

---

[5] We say 'largely' because, although it correctly predicts the abundance of helium-4, helium-3 and deuterium, the prediction for lithium is slightly off. Cosmologists, thinking long and hard about this problem, have decided to name it 'The Lithium Problem'.

helium-3 (deuterium + one more proton) and helium-4 (helium-3 + another neutron).[6]

As we've seen, these processes depend on the temperature of the Universe. The expansion and cooling of the Universe is controlled by the force of gravity. In particular, gravity dictates how much time elapses between the time when the Universe is too cool to make more neutrons and the time when neutrons are bound in nuclei. If this time is more than about 15 minutes, then the neutron stocks will be depleted.

It shouldn't be surprising, then, that changing the strengths of the fundamental forces can have a dramatic effect on the nuclear reactions in the very early Universe. Some of these effects aren't necessarily bad for life. For example, if there were no nuclear reactions at all in the early Universe, then we'd be left with pure hydrogen (remember, the hydrogen nucleus is just a single proton). This isn't such a bad thing: we'd still have galaxies and stars, and so long as stars still worked, the heavier elements would be made there. We would, interestingly, be missing a crucial piece of evidence for the Big Bang theory.

On the other hand, making nuclear reactions too efficient is bad news. All the hydrogen would be burned up in the early Universe. There would be none left over to power stars, and so our stars would run on less-efficient fuel, like helium. Unless the Universe found a way to create some other isotope of hydrogen, such as deuterium (a nucleus with one proton and one neutron) or tritium (one proton and two neutrons), the Universe would not contain water, or any carbon-based molecule that involves hydrogen – which is the vast majority of carbon compounds.[7]

How much would the properties of the Universe need to change in order to give us a helium universe? Consider strengthening the grip

---

[6] There are many nuclear reactions underway in these hectic few minutes of the Universe; the interested reader can play with the most important 87 reactions using the AlterBBN computer code, which can be downloaded from superiso.in2p3.fr/relic/alterbbn/

[7] Of the 24 million carbon compounds listed in the chemspider.com database, fewer than 8,000 (0.03%) do not contain hydrogen.

of the strong force, so that the cosmic oven becomes more efficient. Nuclei can stick together under hotter conditions, and so form earlier in the history of the Universe, when there are more neutrons. Our Universe burns 25 per cent of its hydrogen in the first few minutes. An increase of the strong force by a factor of about 2 is enough to cause the Universe to burn more than 90 per cent of its hydrogen.[8]

If the weak force were weaker, then – perhaps counterintuitively – we would have more neutrons. In the very early Universe, when there is plenty of energy around, the weak interactions are just as happy to make the lighter proton as the slightly heavier neutron. If the weak force is weaker, then at the time when these reactions become inefficient and the Universe ceases to create protons and neutrons (known as *freeze-out*), there are roughly equal numbers of both. So, when the Universe comes to create nuclei by pairing neutrons and protons into a helium nucleus, it can do so very efficiently, locking up all the hydrogen (remember that a hydrogen nucleus is a single proton).[9]

Gravity, too, is implicated in all this. Nuclear reactions don't control the thermostat, because they don't add much to the already blazing inferno of the early Universe. Rather, gravity dictates how fast the Universe cools by controlling the expansion of the Universe. Stronger gravity has the same effect as a weaker weak force – neutron production ends earlier, with roughly equal numbers of neutrons and protons. Almost all of the protons, as before, get locked away inside helium.

The other important factors in this story are the masses of the proton and neutron. Because the neutron is heavier (see Chapter 2), the proton is stable and more abundant. If the two particles had similar masses, or if the neutron were lighter, then we would have more

---

[8] MacDonald and Mullan (2009). In their notation, $G = 1.5$ implies $\alpha_s \approx 2.25\alpha_{s,0}$.

[9] By a 'weaker weak force', we mean a larger weak scale (or Higgs vacuum expectation value), with all other masses and forces held constant. For more details, and a more precise calculation of the effect on Big Bang nucleosynthesis of changing the weak scale and the fundamental particle masses, see Hall, Pinner and Ruderman (2014).

neutrons in the early Universe, and so once again we'd lose our hydrogen.

There are three factors that contribute to the total mass of the proton and neutron. There is the binding energy associated with the strong force, which is the same for both. If that were the only factor, they would have the same mass, so again, no hydrogen – bad news! There is also a contribution from electromagnetism: the proton has to haul some charge around, with its associated electric field. This would make the proton heavier – more bad news. Finally, the quarks have different masses. While they are less than one per cent of the total proton mass, the down quark is sufficiently heavier than the up quark to make the neutron (down-down-up) heavier than the proton (up-up-down) – good news. Once again, the quark masses save the day.[10]

Decay, Radioactivity and the Stability of Matter

Just like the forging of elements at the birth of the Universe and in the heart of stars, the stability of matter is a constant battle of the forces.

Nature has furnished us with 92 natural elements, from hydrogen to uranium, but in the lab we have been able to artificially create heavier elements. Their very short lives are spent seething and wobbling, until the grip of the strong force is broken by the electromagnetic repulsion of the positively charged protons, ripping the nucleus into smaller pieces. The most massive element we have so far created, ununoctium, has a nucleus consisting of 118 protons and 175 neutrons! It tears itself apart into livermorium and helium within about 0.89 thousandths of a second. The livermorium nucleus, in turn, exists for mere moments before it too decays.

But even some of the 92 natural elements are not completely stable. They are *radioactive*, transmuting one element into another, turning lizards into Godzillas, humble country folk into rampaging

---

[10] The effect of changing electromagnetism is slightly complicated. You might think that weakening the repulsion would make nuclear burning easier and more efficient. In fact, the dominant effect is on the neutron–proton mass difference, with stronger electromagnetism meaning heavier protons and thus more neutrons are produced. The result is a universe dominated by helium.

flesh-eating mutants, and mild-mannered teenagers into superheros and supervillains. (We watch too many movies!)

In reality, radioactivity is both a curse and a blessing, with whole branches of medicine dedicated to using the decay of radioactive atoms to identify and kill cancerous tumours. When you have a moment, do a little research on how Positron Emission Tomography (PET) or Single Photon Emission Computed Tomography (SPECT) scanning work.

Radioactivity is a complicated beast, a catch-all for several different processes. In the most general sense, radioactivity is the spontaneous emission of energy from a nucleus. These energetic products can destroy or illuminate living tissue.

Different types of radioactivity can be classified by the particle emitted. The simplest is perhaps gamma(γ)-decay, where an excited nucleus emits a high-energy photon, in much the same way that the electrons of an atom can emit photons as they jump between energy levels. Unlike other decays, the emission of the photon in γ-decay does not change the number of protons and neutrons in the nucleus, and so the elemental nature of the atom does not change.

Alpha(α)-decay occurs when a heavy element emits the nucleus of a helium atom (two protons and two neutrons), which in this case is called an α-particle. By removing protons, α-decay transmutes one element into another.

The process of α-decay is a battle of forces: the short-range strong force, which holds protons and neutrons together in the nucleus, and the electromagnetic repulsion of the positively charged protons, which would tear the nucleus apart.

We can imagine that the nucleus is like a prison yard, holding the α-particle inside.[11] It can roam freely inside the yard, occasionally bouncing off the walls. This bouncing is like the action of the strong force, preventing escape.

[11] This is a surprisingly apt metaphor. See Cohen (2008) for a simple mathematical model of the stability of nuclei using square potential wells.

If we calculate the energy of the α-particle, and compare it to the energy needed to jump the prison walls, we will conclude that escape is impossible. The particle can, however, cheat: the rules of quantum mechanics allow for the possibility that the particle will *tunnel* to the outside without ever having had the energy to jump over the wall.

The α-particle within a nucleus may try to escape its prison many trillions of times per second. For a slowly decaying nucleus, the vast majority of these attempts will be unsuccessful. In a typical uranium-238 nucleus, for example, the α-particle will bang its head against the wall a trillion trillion trillion times over a few billion years before successfully tunnelling to the outside world. Once it finds itself beyond the powerful but short reach of the strong force, electromagnetic repulsion drives it rapidly away from the nucleus.

This quantum tunnelling sounds like a fudge, something that quantum physicists made up at a party after a few too many Chardonnays. But it's a real phenomenon and is regularly observed in the laboratory. It occurs in modern electronics, too: if wires are packed so close together that electrons can tunnel from one to another, electrical signals will be disrupted.

Finally beta(β)-decay occurs when a nucleus spits out an electron, or its antimatter cousin, the positron. This form of radioactivity is a manifestation of the weak force.

Recall that the weak force can *transmute* fundamental particles. At the W^+ and W^- vertices, quarks can be interchanged and a member of the electron family can be swapped for one of the neutrinos. To see the weak force in action, we'll take a closer look at the decay of a neutron.

It may seem strange that the neutron, one of the key building blocks of all atoms and responsible for about half of your mass, is unstable. If you take one outside of a nucleus and leave it alone for about 15 minutes, it becomes a proton, releasing an electron and an antineutrino in the process. Inside a nucleus, thankfully, the neutron can be stable.

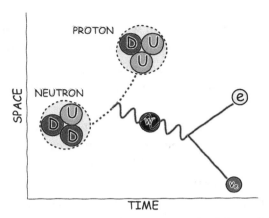

FIGURE 18 Neutron decay: an example of the weak force in action, with a neutron emitting a W⁻ particle as it changes into a proton. The W⁻ then decays into an electron and a neutrino.

We can understand the decay of the neutron in terms of the weak force. As shown in Figure 18, one of the neutron's two down quarks becomes an up quark. The neutron has become a proton, and a W⁻ particle is ejected; notice that the total charge does not change. The W⁻ is short-lived, travelling a tiny distance before decaying into an electron and its ghostly companion, the electron-neutrino. The electron is usually too energetic to be bound by electromagnetism into an atomic orbit around the proton, and so escapes completely.

In certain nuclei, these weak processes change neutrons into protons, with the emission of an electron, or even flipping a proton into a neutron, spitting out a positron. With each β-decay, the nucleus switches from one element to another.

Radioactive, More or Less

As we have seen, radioactivity is a result of the combination of the fundamental forces. In α-decay, a helium nucleus escapes the clutches of the strong force, while β-decay is the weak force in action, flipping protons into neutrons and neutrons into protons.

Gamma radiation consists of high-energy photons, emitted by an excited nucleus.

The rate of α-decay depends upon the balance between the pull of the strong force and push of the electromagnetic force within the nucleus. What matters to quantum tunnelling is the *thickness* of the prison wall, that is, how far does an α-particle need to tunnel before repulsion wins over attraction?

Similarly, the rate of β-emission can be changed by playing with the strength of the weak force, modifying the coupling constant that controls the probability of the emission of the weak force (gauge) bosons, the Z^0 and $W^{\pm}$.

Let's imagine what a radioactivity-free universe might look like. Perhaps, apart from missing out on a few superheroes, supervillains and associated monsters, the universe would be a safer place. However, eliminating radioactivity has some surprising consequences.

At the Earth's Core!

Solid ground feels pretty solid. It comes as a surprise to realize that the Earth's surface is merely a thin crust enveloping a hot interior of rock and iron. How thin? Picture the skin on an apple – *that* thin.

Below the surface, on the immense timescales of geology, rocks flow, dragging continents, raising mountains and unleashing volcanos. Earth's hot liquid rock layers generate its magnetic field, which funnels charged particles streaming from the Sun, known as the *solar wind*, towards the magnetic north and south poles, producing beautiful aurorae and protecting life on Earth from this harmful radiation.

The Earth was born as a hot ball of rock, and has been cooling over the 4.5 billion years of its existence. Without an internal source of heat, the centre of the Earth would have cooled and solidified long ago. However, energy is injected into the rocks through the continual radioactive decay of its elements, particularly isotopes of potassium, uranium and thorium.

Today, the radioactive heating of the Earth is dominated by α-emission by long-lived uranium and thorium atoms, while in the past

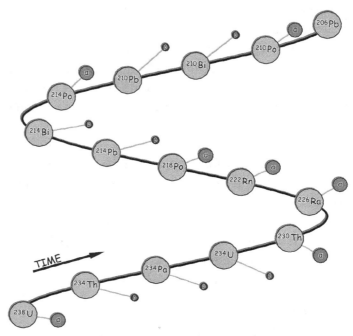

FIGURE 19 The radioactive decay from uranium to lead proceeds through
a series of steps, each emitting either an α- or a β-particle at each stage.

the dominant heating was through β-decay of potassium. Clearly, the
strengths of both the strong and weak forces will influence the total
amount of radioactive energy released into the Earth.

Let's take the decay of uranium as an example. Within the Earth,
the main uranium contribution to heating comes from the ^{238}U isotope,
which has an extremely long half-life of almost 5 billion years. Thus,
there is uranium in the Earth today that was there when the Earth was
born. After the emission of an α-particle, we are left with a nucleus of
thorium (^{234}Th to be exact), which is also unstable and decays in less
than 25 days through β-emission. This series of α- and β-decays continues
through 27 steps, via elements with half-lives ranging from seconds to
aeons, until we arrive at lead, a stable nucleus (Figure 19). Each step of
the decay injects energy into rocks inside the Earth.

Without this energy keeping the Earth warm, it would have solidified, losing its protective magnetic field. Our planetary neighbours, Mercury, Venus and Mars, have virtually no magnetic field and so are blasted by the solar wind, boiling elements off their atmospheres.

Also, without heating by radioactivity, the Earth would be solid and so without plate tectonics. At first glance, this seems to be a great improvement: no shifting plates so earthquakes, volcanos and tsunamis are a rarity. There are, however, serious downsides. The constant churning and upheaval as plates collide and volcanos erupt plays an important role in keeping nutrients cycling through the biosphere. So, the rich soils of Java, Japan and Hawaii are due to volcanic activity, and the slow uplift of the Earth's crust that creates mountain ranges has also brought fertile agriculture to parts of North America, India and Europe. By contrast, Australia's stable tectonic plates and few volcanos explain why its soils are relatively unproductive: they are old and leached of nutrients by billions of years of rain.[12]

The cycling of carbon through the crust and the atmosphere stabilized the Earth's temperature, hovering between the extremes of frozen oceans and the runaway greenhouse effect that stifles Venus. Furthermore, the changing face of planet Earth leads to the emergence of new environments and ecological niches, thought to be essential for the evolution of complex life.

Obviously, too much radioactivity in the Earth's core is no good either. If radioactivity had maintained the Earth's initially molten surface, then the prospects for complex, intelligent life swimming in a molten sea of rock would have been very limited indeed. No worries, you may think, as the energy of radioactivity is finite. After most of the nuclei have decayed, the Earth can cool. Unfortunately, the Sun's energy is also finite: it would be a tad pointless for the Earth's surface to solidify just in time to be engulfed by the Sun, as it exhausts its hydrogen fuel and swells to become a *red giant star*.

---

[12] See Diamond (2005, Chapter 13) for more details.

INTO THE VALLEY OF STABILITY

One piece of the nuclear puzzle remains to be put in place: which nuclei are stable? A chemical element is defined by the number of protons in the nucleus, known in the trade as the *atomic number*. The chemical properties are then determined by the distribution of electrons orbiting the nucleus; for a neutral atom, there will be the same number of electrons as protons. So, every atom of lead contains precisely 82 protons, while every atom of nitrogen contains 7.

Nuclei also possess a certain number of uncharged neutrons. These attract each other and protons via the strong force without adding electromagnetic repulsion, and so help to stabilize the nucleus. Two nuclei of the same element (that is, with the same number of protons) can have different numbers of neutrons. For lead, some nuclei may have as few as 96 neutrons, or 97, or 98, all the way up to 133 neutrons. Each of these *isotopes* of an element has a different weight.

We've seen that the most massive elements spontaneously fall apart, their internal forces unable to hold them together against the nucleus's constant shuddering and pulsing. Even less massive elements can transmute into other elements through the action of radioactivity. Of the 2,000 isotopes known, less than 300 are stable. This list is revised every so often when it's discovered that a 'stable' isotope actually decays with an extremely long lifetime.

Why can't we keep adding more and more neutrons, and so build more and more massive nuclei of a particular element? Remember that quantum things aren't very good at sitting still. The nucleus pulses and surges, spins and deforms. Neutrons are fermions, and you'll remember from Chapter 2 that identical fermions don't like to be packed together. The *dineutron* – two neutrons trying to stick together – is unstable for this very reason, and falls apart into two neutrons in 10^{-22} seconds. So, in a nucleus with too many neutrons, some will be too loosely bound and so rapidly lost.

Similarly, adding too few neutrons is no good either. While the proton is stable, the *diproton* is not: we can't stick two protons together. Not only do these fermions object to being in each other's space, their positive charges repel. Like the dineutron, it will fall apart. Nuclei with too many protons are liable to lose one or a few.

Even nuclei that bind all their protons and neutrons are not safe. They might be susceptible to decay via the interactions of the weak force. If a nucleus were more tightly bound with a proton instead of a neutron or vice versa, then the weak force would make this happen; this is β-decay.

Putting this all together, we can make a picture of all of the isotopes of all of the elements, showing those few that are stable and the broader swath that are unstable. As we make our way through the periodic table, we have to add more and more neutrons to help stabilize the nucleus. The stable nuclei run in a narrow band though the centre, known as the *valley of stability* (Figure 20).

As we have seen, the valley of stability is the result of the interplay between three of the fundamental forces of nature. The strong force holds the protons and neutrons together. Electromagnetic repulsion acts pushing the protons apart. The weak force can turn a proton into a neutron, or vice versa.

Broadening the Valley of Stability

Clearly, messing around with the strengths of these fundamental forces will influence the structure of the valley of stability. A universe with no weak interactions will have no β-decay. The result is that some unstable isotopes would become stable, while others would instead decay via emitting an α-particle. Alternatively, weakening the push of electromagnetism would hold more nuclei tightly together, preventing the escape of α-particles.

Would these additional stable isotopes be beneficial for life? Fans of 1960s war movies will remember *The Heroes of Telemark*,

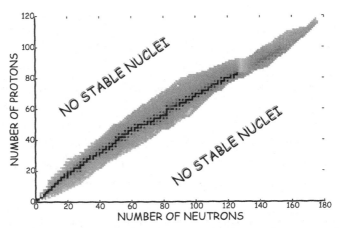

FIGURE 20 The valley of stability. All of the elemental isotopes we know, with the darker shades being more stable and the black totally stable.

in which the dashing Kirk Douglas is on a mission to destroy the Vemork Hydroelectric Plant in Norway. Of course, Douglas and his valiant team succeed and the rest, as they say, is history.[13]

The plant in question produced *heavy water*, which was crucial to German nuclear weapons research. Molecules of heavy water contain two deuterium atoms instead of two hydrogen atoms. Deuterium is an isotope of hydrogen, in which a neutron accompanies a proton in the nucleus. This extra mass makes heavy water 10 per cent heavier than normal water, hence the name.

Chemistry would not change much; the interaction of the outer electrons of atoms of various isotopes of an element would be similar but not identical. For example, we can create heavy water molecules from heavy hydrogen (deuterium). Bonds between heavy water molecules are slightly stronger than those of ordinary water, which is bad for biological systems. So bad, in fact, that heavy water, while it

---

[13] As with all Hollywood movies of real events, the film contains an element of truth. The actual raid, a culmination of several attacks, undertaken in 1943 by Special Operations Executive Norwegian Commandos, was as thrilling as the movie and well worth reading about (such as *The Real Heroes of Telemark* by Ray Mears, 2003).

apparently tastes (at least to humans) like normal water, is actually toxic. In molecular interactions, heavy water just doesn't do what ordinary water does. And this is a problem not only for humans, but for multicellular organisms in general. This probably reflects the adaption of life on Earth to the material available; we should not conclude that life based on heavy water is impossible.

What if we could banish radioactivity altogether? At the moment, such a universe is too complex to be explored in detail. But we can make some informed guesses.

Consider carbon, the cornerstone of life as we know it. Each carbon nucleus has 6 protons, and can take a range of neutrons. Its most massive isotope has 16 neutrons, whereas the lightest has only 2. This means that the heaviest isotope of carbon is almost three times as massive as the lightest! These extremes are at the edge of the valley of stability and, in less than a thousandth of a second, decay away. In fact, there are only a few isotopes of carbon that are stable enough to be important for life, those with 6 or 7 or 8 neutrons.

However, if we flatten the valley of stability, then very light and very heavy isotopes of carbon will be stable. And it is the same for all of the other elements. The result is that the molecules built out of this sea of atoms, such as amino acids and proteins, may not be chemically identical. They could have differing masses, molecular bond strengths and shapes, and hence different functionality.

Could the complex molecular interactions that underpin life arise in such a seemingly complicated mess of isotopes? We don't know. It certainly seems to place another obstacle in the way of chemical evolution, that is, the formation of the first life forms. This is something that those that think about fine-tuning will have to ponder in the future.

Of course, as we have already noted, wiping out radioactivity and flattening the valley of stability has a detrimental effect on planets like Earth. We need that molten core and its magnetic field and plate tectonics. It seems that life, at least life like ours, has benefited from the sides of the valley of stability being unstable!

Erasing Stability

As we saw above, adding more stable elements to the Universe we know and love is unpredictable. We're not really sure what would happen. It is much easier to see fine-tuning when something vital to life is simply *removed* from our Universe. This is a definite step backwards. Erasing the valley of stability entirely is just such an example.

A supervillain who threatens to dissolve our atoms and nuclei is worth taking seriously. Chemical elements that last but a moment are pretty useless for building molecules, cells and organisms and it's not difficult to arrange: decrease the strength of the strong force by a factor of about 4, and the periodic table, from carbon up, is gone. They don't even α-decay. They experience *fission* – the nucleus simply splits into two. We could achieve the same effect by increasing the strength of electromagnetism by a factor of 16. Just don't tell your local Evil Overlord.

Smaller changes will result in most of the elements being radio-active, with a range of lifetimes. This creates a swath of problems for life. Each α- and β-decay transmutes one chemical element into another, and so totally changes its chemical properties. Within life's highly specialized, precisely constructed amino acids, proteins and cells, an unpredictable change of this magnitude will cause untold damage. Life will not get very far if it cannot trust chemistry.

Further, α-, β- and γ-radiation have a deservedly bad reputation. Radioactivity was first discovered in uranium, where it was noticed that a sample would gradually darken photographic film (via X-rays). Then Marie and Pierre Curie discovered radium, which made radio-activity into an overnight sensation: this stuff glowed in the dark, a seemingly inexhaustible source of energy. It was soon being sold in all kinds of products to increase your metabolism, vitality and general wellbeing! Products were advertised as containing the power of radium – why not visit the radium water baths, drink some radium water, buy radium boot polish and radium starch ... even radium condoms! (Thankfully, many of these products don't seem to have contained any actual radium; a rather fortunate case of false advertising.)

It wasn't long before it was realized that this was a very bad idea. *The Wall Street Journal* ran an article with the extraordinary headline: 'The Radium Water Worked Fine Until His Jaw Came Off'.

We now know why. When α- and β-decaying nuclei enter the body, they release high-energy particles inside cells. Life's molecules don't stand a chance, and a trail of destruction is carved through molecular machines, cell walls, and – most destructively – DNA. A typical α-particle has enough energy to break tens of thousands of chemical bonds. While our cells have self-repair mechanisms, and a small amount of genetic mutation is necessary for Darwinian evolution, there is only so much life can handle. A universe in which any of the six elements common to all life – carbon, hydrogen, nitrogen, oxygen, phosphorus and sulphur – were radioactive would be a dangerous place for life.

So, the properties of the valley of stability, properties defined by the relative strengths of the strong, weak and electromagnetic forces, appear to be nicely balanced for life in the Universe. We have a narrow floor of stable isotopes to provide the solid and stable molecular structures needed for life, with a broader swath of longer-lived, but ultimately unstable, isotopes that can provide radioactive heating within planets like the Earth. In universes with differing strengths of these fundamental forces, the prospects for life seem at least more complicated, if not totally eliminated.

JUST ONE FORCE FEWER?

What if we were really radical and removed a fundamental force from the Universe entirely? We could do this by setting the coupling that defines the strength of a force to zero. This is generally a bad idea if you want to create a hospitable universe.

Turn off gravity, and there is nothing to drive matter to collapse into galaxies, stars and planets, or indeed any structure. Turn off electromagnetism, and there is no chemistry as there is nothing to keep electrons bound to nuclei. Turn off the strong force, and there are no nuclei, and so again chemistry is doomed. We have little choice but to pick on what may seem like the most inconsequential force. Could we get away with turning off the weak force?

Radioactivity has a crucial role in the generation of heat within the Earth, but even if we completely extinguished the weak force, we could still have α-decay of elements contributing to radioactive heating. But, as we have seen, the chain of radioactive decays that deposits the energy into the rocks is a mix of α- and β-decays. Without the weak force, these chains would be cut short, reducing the heating within the planet.

Roni Harnik, Graham Kribs and Gilad Perez have investigated this *weak-less universe*, to see if life has a chance.[14] Let's have a closer look.

One can imagine the weak force being turned off by ramping up the Higgs field so that the masses of the Z^0 and $W^\pm$ particles are huge. All weak processes create these particles, even if fleetingly, and so if they are extremely massive then they will be created very rarely. The weak force all but disappears.

But wait ... if we've made anything that gets its mass from the Higgs field extremely massive, wouldn't that make the quarks and electron similarly enormous? Not quite. We can simply stipulate that particles, fortuitously, couple very weakly to the Higgs field. Obviously, this makes the weak-less universe fine-tuned anyway, but let's continue the tour.

Without the weak interaction, neutrons would be stable,[15] and all protons and neutrons would be forged into helium, leaving behind

---

[14] Harnik, Kribs and Perez (2006).

[15] In fact, it isn't clear that there will be any matter in the weak-less universe at all, as the weak interaction appears to be crucial for creating more matter than antimatter. Without this asymmetry, almost all matter annihilates into photons in the very early universe. We will come back to this in Chapter 6.

the merest trace of hydrogen. As we noted above, this is bad news for stars, water and almost all known organic molecules.

The weak-less universe can be saved, however, with a little 'judicious parameter adjustment', as Harnik and co. call it. We could write the desired neutron to proton ratio (via the quark ratios) into the initial conditions of the universe. With an excess of protons, there are leftovers when the universe burns two protons and two neutrons into helium. As we are unsure of the mechanism that produces this initial property of the universe, we can happily suppose that it too could have been different.

Alternatively, and perhaps less suspiciously, we could increase the number of photons in the universe by a factor of 100. This makes the universe more efficient at destroying newly formed nuclei, destroying some of that hydrogen-devouring helium.

Such universes may offer alternative approaches to fuelling stars and creating heavier elements, potentially resulting in complex chemistry. However, there is another worry. To date, the most detailed look at the production of elements in stars and supernovae has concluded that removing the weak force results in an oxygen-deprived, and thus water-deprived, cosmos.

This weak-less universe is ingenious, and the universe-constructing exploits of Harnik and co. make for entertaining reading. Removing the weak force significantly changes the face of the universe, but is somewhat surprisingly not necessarily a death sentence for life. The weak-less universe, as we noted, is nevertheless fine-tuned.

Interestingly, the occupants of the weak-less universe would have a very difficult time uncovering their fundamental physics. Discovering the physics of the weak force, including the Higgs field and the origin of the fundamental particle masses, would require probing extremely high energies, a thousand trillion times beyond our (and almost certainly their) largest particle accelerators. But even if they managed to determine how physics worked, they would be even more baffled by their apparently fortuitous cosmos than we are.

We've come to the end of our discussions of forces. It's time to talk about the other principles that animate life in the Universe: energy and entropy.

4   Energy and Entropy[1]

The Universe runs on energy. This energy comes in a variety of forms, such as heat, light, motion, and in matter itself, which Einstein's famous $E = mc^2$ tells us is also a form of energy.

The flow of energy drives life's countless processes. However, any old energy will not do; life needs energy in a useful form. This leads us to one of the most important and challenging, and most widely misunderstood, concepts in the physical sciences: *entropy*.

In this chapter, we will trace the flow of energy and entropy through the cosmic ages. We will see that the Universe is inexorably winding down, with less and less useful energy available. This holds grave concerns for the future, and points to one of the great mysteries of modern science: why was it born in a wound-up state in the first place?

In the preceding chapter, we investigated the fundamental forces, mostly focused on the strong, weak and electromagnetic forces. The elephant in the room is gravity. We have saved gravity until this chapter, as it dictates the cosmic flow of energy; though the weakest of all the forces, it has a starring role in shaping the Universe.

Let's start with humankind's single-minded quest for energy.

HUMANS NEED ENERGY!

Humans are obsessed with energy. Our need for energy has altered the very surface of Earth. Looking at the night side of Earth from space shows its surface to be dotted and criss-crossed by the lights of cities and highways that cover much of the planet.

---

[1] We like to imagine Paul McCartney and Stevie Wonder singing this.

Individually, our bodies need to be supplied with energy, and regularly tell us so. We crave a burger with all the trimmings, a dodgy pizza, or fine Belgian chocolate with a nice French brandy.

Humankind's search for energy, in the form of food, has shaped our evolution, the development of civilization, and now most television. After a few hours without food, the pangs begin. After about a month without food, your body simply can't take the strain anymore, and you kick the bucket.

You're an energy-processing machine. All life is.

Like many human endeavours, the study of energy – in particular the flow of heat – was driven by money. In the eighteenth and nineteenth century, the industrial revolution rode on the back of the steam engine. More efficient engines meant more money, and so inventors and scientists were drawn to investigate how heat can be turned into motion. From the factory floor came the science of *thermodynamics*.

The importance of thermodynamics to modern physics can scarcely be overstated. It provides the guiding principles and mathematical framework to understand any energy flow in the Universe, from the digestion of a chicken nugget, to the powering of the leg muscles of a rugby player, to the nuclear reactions that power the sunlight that fuels the plant that makes the seed that nourishes the chicken that becomes the nugget.

There are four fundamental laws of thermodynamics, but for historical reasons they are numbered zero to three. The cynic's version goes like this. Zero: there is a game. One: the best you can do is break even. Two: You can only break even at a temperature of absolute zero. Three: You can't reach absolute zero.

More precisely, and perhaps less depressingly, the laws are as follows. The zeroth law allows us to define the concept of *temperature*. If two objects have the same temperature, then heat will not flow between them; they are in *thermal equilibrium*. The first law says that energy (including heat) is conserved, that is, neither created nor destroyed. The third law states that *absolute zero*, the temperature where all thermal motion *theoretically* ceases, cannot be reached in practice.

Here we will focus upon the most famous of the laws of thermodynamics, the second law. To underscore its reputation, here's the great British astrophysicist Arthur Eddington (1928, p. 74):

> If someone points out to you that your pet theory of the universe is in disagreement with Maxwell's equations [of classical electromagnetism] – then so much the worse for Maxwell's equations. If it is found to be contradicted by observation – well, these experimentalists do bungle things sometimes. But if your theory is found to be against the second law of thermodynamics I can give you no hope; there is nothing for it but to collapse in deepest humiliation.

The second law of thermodynamics (roughly) states that the amount of order in the Universe (or isolated part thereof) does not spontaneously increase. Order is at best maintained, and only created at the expense of disorder elsewhere. The term *order* is intuitive but a bit fluffy; physicists prefer to talk about something called *entropy*.

Entropy, Disorder and the Energy of the Universe

We're familiar with order descending into disorder, and with the effort needed to restore order. That kitchen won't clean itself. Gluing the broken vase back together is going to be a lot harder than breaking it in the first place. The ice in your glass will melt and the Scotch will reach room temperature; you've never seen that process go backwards.

However, entropy's importance as a concept is matched by its subtlety. Just ask a physics undergraduate who has completed a course on thermal and statistical physics. Entropy is easy to misunderstand.

We can simplify things if we think of entropy in terms of *useful* or *free energy*. A system with low entropy has energy that can be extracted and converted into another form. By contrast, the energy of high-entropy systems is stuck, unable to be tapped.

Consider a hydroelectric power plant. The water in the mountains has useful energy, in this case, *gravitational potential energy*. As it flows down the mountain, it can turn a turbine, allowing this energy to be tapped and extracted to do useful work. When the water has reached the lake at the valley floor, it has been robbed of its potential energy, and can no longer be used to drive the turbine. In the mountains, the water was a low-entropy energy source, but when it reaches the valley lake, its entropy has increased.

Let's continue with a simple example about heat. Consider two buckets of water, one hot and one cold. The system consisting of these two buckets has low entropy, so you should be able to do something useful with them. For example, you could connect the two buckets with a metal wire. Heat will *flow* from the hot bucket to the cold. With the right materials and a bit of cleverness, you could build a *thermoelectric generator*, which makes use of this flow of heat to produce electricity.

As heat flows between the buckets, the hot one gets cooler and the cold one warmer. The water in the buckets will eventually reach the same temperature, at which point everything stops. No heat flows, and no electricity is produced. There is no more useful energy to be extracted. The two buckets are now a high-entropy system.

Note that there is still plenty of thermal energy. It just isn't going anywhere, or changing form.

But what about 'order'? Is the initial state of one hot bucket and one cold bucket more ordered than the final tepid state? We can think of it like this. Start again with a hot bucket and a cold bucket, and pour all the water from the hot bucket into 50 glasses, and all the water from the cold bucket into another 50 glasses. Of the 100 glasses, choose 50 at random to pour back in the first bucket, and 50 into the second. There are 10^{29} ways to do this, but only *two* restore the hot and cold buckets (one for the original buckets, and one that swaps hot and cold). Every other way mixes hot and cold, reducing the useful temperature difference. While we have not changed the total amount of thermal energy, the vast majority of ways produce two tepid buckets of roughly the same temperature, with little useful energy.

Thus, the initial system, with hot water over there and cold water over here, is far more ordered than the final system with tepid water in both buckets. Entropy has increased as the buckets have come into equilibrium.

We could restore the initial state of the system by heating one bucket and putting the other in a refrigerator. The buckets would be useful again, able to generate electricity. But, when we settle the accounts, squaring the ledger of all the energy used to heat and cool the buckets, we invariably find that the total entropy has increased. The restored low entropy of the buckets has come at the expense of disorder elsewhere.

The second law of thermodynamics states that in any situation where energy flows, the total entropy never decreases. At best, the total entropy remains the same. More often, the entropy will increase, decreasing the energy available to do other things.

In our expanding Universe, entropy is steadily increasing, and the amount of available energy to do things is constantly decreasing. In the distant future of the Universe, the available energy will dwindle away, leaving it bland and lifeless, a little bit like Canberra.[2] We will return to this later (the Universe, that is, not Canberra).

Life, whatever its defining features, is characterized by action. Life is driven by energy processing, and the concepts that dictate the flow of energy between our hot and cold buckets also apply to you. No energy, no life. But, like our buckets, energy alone is not enough to create electricity or motion or metabolism. We need *useful* energy. Life needs low entropy.

So, where does life get its useful energy? Let's trace our supply of useful energy back in time.

---

[2] For our non-Australian readers, Canberra was selected as the national capital in 1908 to prevent the bickering between the cities of Melbourne and Sydney, a bickering that continues to this day. Like many planned cities around the world, it has a much-maligned reputation as being boring and devoid of nightlife. Of course, it is not really that bad (but it is not as nice as Sydney!).

The March of Entropy

The ultimate source for (virtually) all of the energy on Earth is the Sun. At the heart of the Sun, the immense temperatures and pressures allow hydrogen nuclei to be squeezed together to produce helium. This process releases energy, which keeps the nuclear furnace burning. However, as the core accumulates more and more helium, the available energy decreases and the entropy increases.

While the Sun will be able to liberate energy from this helium core in the future, fusing lighter nuclei to make heavier elements, this process cannot continue forever. Eventually, the atoms at the heart of the Sun will have nothing more to give. The fires will go out and the star will die. Humanity, hopefully, will have moved elsewhere.

But let's return to our present-day Sun, happily burning hydrogen. The energy it releases streams through space, and a very small portion of it lands on planet Earth. Much of it lands on water, rocks and sand, but some is absorbed by plants and algae.

Through *photosynthesis*, plants use the energy in sunlight to convert simple chemicals into more complex chemicals, storing energy in the form of chemical bonds.[3] The best photosynthesis can do is lock away as much sunlight as it receives in chemical energy, but a consequence of the second law is that some energy will always be wasted, in this case lost as heat. Again, the useful energy is diminished, and the entropy continues to rise.

The energy stored by plants is used to build more plant. Animals can take advantage of the plant's hard work by eating it, gaining its stored energy. Of course, as omnivores, humans often (literally) cut to the chase and consume other animals, who have stored the plant's energy in their muscles, organs and fats.

What becomes of the energy we consume? Even sitting in bed, doing nothing more than reading a good book (like this one), you are usually warmer than your surroundings. Your inner heat warms the

---

[3] In keeping with a running theme of this book, reality is slightly more complicated than our discussion.

air around you, and is radiated away as infrared rays; if your eyes were sensitive to this part of the electromagnetic spectrum, people would glow. Our processing of food is nowhere near 100 per cent efficient, and our waste products go on to fuel plenty of life in the sewers and beyond.

Modern civilization is powered not only by the plants and animals we consume, but also by the oil, gas and coal we dig out from the ground. What is the source of this large reservoir of energy?

Geraint is the son of a coal miner. His father often brought home pieces of coal that, when split open, revealed a beautiful imprint of an ancient fern. This helped spark his son's interest in science.

These ferns lived more than 70 million years before the first dinosaurs, at a time when giant insects roamed the land and air, and sharks dominated the sea. During this *Carboniferous era*, large swathes of the Earth were tropical rain forest and swampland, with countless plants capturing sunlight.

When some of these plants died, they didn't have a chance to decay, as their remains were locked away under airless mud. As their material compacted under layers of sediment and rock, through immense pressures and heat, this plant material became coal. At other times, algae and tiny marine animals became trapped within the rocks, slowly transforming into oil. It is the bodies of these long-dead plants and animals we extract today. The oil in your car and the coal that generates electricity are little more than condensed sunlight.

Just like the hot and cold buckets from the previous section, useful energy flows from the hot Sun to the cooler Earth in the form of sunlight. This stream of photons is able to power all sorts of useful processes on Earth, and so it must have relatively low entropy.

The energy is processed, some by life, and for every high-energy photon that the Sun sends to Earth, about 20 low-energy photons are radiated into space.

The entire enterprise of your existence – using energy to turn food into more you and more descendants – draws on the bank of low-entropy radiation streaming from the Sun. *Some* source of low entropy is needed for any activity in the Universe, including life. So, why is the Sun a source of low entropy?

THE WEIGHTY MATTER OF GRAVITY

As we've seen, our everyday lives are a battle between fundamental forces. Gravity is pulling us towards the centre of the Earth. The Earth's atoms – thanks to electromagnetism – resist being squeezed, and so support our weight. Given gravity's grip of everything around us, it may come as a surprise just how weak gravity is compared to the other forces. Consider two protons: the ratio of their gravitational attraction to their electromagnetic repulsion is about 10^{-36}! Electromagnetism is trillions of trillions of trillions of times stronger.

If you are unsure about the weakness of gravity, imagine a bowling ball falling from a skyscraper. Pulled by the Earth's enormous mass (10^{24} kg), the ball will fall, accelerating towards the centre of the planet. Neglecting air resistance, for every second the ball falls, it gains another 9.8 m/s in speed. It plummets faster and faster, until it hits the floor. Then what happens? Well, let's consider a very hard floor, made of steel. And let's also assume the bowling ball is solid steel, so it won't crack or break. Remember, gravity has pulled on the ball for tens or hundreds of metres. But the electromagnetic force, provided by the repulsion of the outer electrons of the atoms in the floor and those in the ball, sends it bouncing back in the space of a millimetre. While gravity is important to us, there is no denying that it is puny.

If you've been paying attention, you may be wondering: if gravity is so weak, why doesn't electromagnetism completely dominate the workings of the Universe? There is a crucial difference between electromagnetism and gravity. There are two kinds of electric charge, positive and negative, which can cancel each other out. By contrast, there is only one gravitational 'charge' (i.e. mass), which tries to accumulate more mass. The electromagnetic interaction that holds an atom together is almost invisible to the outside world because the number of negative charges (electrons) balances the number of positive charges (protons). At a distance, there appears to be no charge at all. As we make larger and larger things out of equal numbers of protons and electrons, with charge cancelling out and mass adding up, gravity will eventually win.

So gravity, while the weakest of the forces, plays a vital role in our very existence. Gravity shapes our cosmic environment. Let's take a look at its role in creating sunlight.

Set Controls for the Heart of the Sun!

We have seen that the Sun is the dominant source of energy powering the Earth. Whether this energy arrived millions of years ago and is now stored as coal and oil, or arrived recently and was converted into electricity via a solar panel, or through photosynthesis into plant material, this is the energy that allows you to exist as a functioning human being. The ultimate source of its energy, however, remained a mystery for a long time.

Only in the last one hundred years have we come to understand that the Sun is an immense pressure cooker. With its extreme mass, more than a million times that of the Earth, the Sun is able to squeeze its core to immense pressures and temperatures. With energetic protons pushed tightly together, nuclear reactions are ignited, creating helium. Energy is released as very high frequency photons, known as gamma rays. Such high-energy rays are destined to become the sunshine falling on the Earth, but in this raw state such gamma rays would either bypass or destroy the molecular structures inside us.

These newly formed photons, however, do not have an easy path from the core of the Sun to the surface of the Earth. In fact, given the extreme densities, they start bouncing off the surrounding protons and electrons, heating the gas. Thanks to this extremely high temperature, the gas's outward pressure balances the gravitational squeezing.

Just as the jolly drunkard staggers from the bar, these photons make their way out from the core of the Sun, bouncing countless times, and eventually reach the solar surface. It takes hundreds of thousands of years to cover a distance that light could cover in a few seconds, if only it could travel in a straight line. Released at the solar surface the photons stream through space to give us spectacular sunrises and sunsets, and to fuel our lives. Their bouncing path through the Sun has important consequences. As well as supporting the gas against gravity, it has dialled down the energy of the photons from sterilizing gamma rays to plant-fuelling optical photons.

Most stars can live in this happy equilibrium between the inward squeezing of gravity and the outward pressure force for billions of years. If the core is squeezed too much by gravity, it will push back. If it expands a bit too much, gravity will squash it back into place.

However, some stars that have exhausted the hydrogen in their cores have nuclear burning rates that are sensitive and fickle, and which are unable to establish this balance. When squeezed, the core over-reacts: a burst of energy is released into the star that causes it to expand too far. The star stalls and contracts, causing another overreaction. Rather than settling down, these *Cepheid variable* stars pulse.

For some stars, the pulsing becomes increasingly violent, until there is one massive palpitation that drives off the outer layers of the star into space, leaving nothing behind but a dying stellar core. This sobering scenario is the eventual fate of the Sun, roughly five billion years from now.

As we have seen, the flow of energy from the heart of the Sun ultimately powers life on Earth. Before the first stars ignited, there were no low-entropy sources of light in the Universe that life could tap. The birth of the first stars is an important step in creating life in this Universe.[4]

---

[4] Harvard astrophysicist Avi Loeb (2014) has speculated that life could have arisen in the warm bath of radiation left over from the Big Bang. This radiation, which was originally ultra-energetic gamma rays, cools as the Universe expands. About 10 to 20 million years after the Big Bang, the Universe is warm, being about the same temperature as Earth. Rocky planets – if there are any – could hold liquid water.

Is this an environment for life? Remember that it is temperature *differences* that allow you to extract energy and drive life, and this cosmic background radiation is an almost uniform sea, with no flow from warm to cold, so powering life is very problematic.

Stars are powered by nuclear *fusion*, combining lighter elements to create more massive nuclei. As with the early Universe, this is a battle between all four fundamental forces. The electromagnetic force pushes the positively charged particles away from each other, while the strong force provides the glue to join the lighter nuclei into heavier elements, liberating energy.

The weak force, too, plays an easily overlooked but crucial role. You may have noticed that stars burn hydrogen nuclei (each consisting of one proton) into helium (two protons and two neutrons). Where did the neutrons come from? During the collision of two protons, it is possible (though unlikely) that one will transform into a neutron thanks to the weak force. Given the sheer number of proton collisions in a star, even these rare events can become very important. The proton and neutron together make a deuterium (heavy hydrogen) nucleus, and a positron and neutrino are emitted from the reaction. A neutron is made, and can be used to make helium.

There is a crucial difference between gravity's role in stars and in the cosmic oven. In the first few minutes of the Universe, gravity's cooling effect doesn't much care whether nuclear reactions are taking place. The Universe will expand and cool regardless. Thus, the nuclear reactions have one and only one chance: there is one short period of time when the Universe has the right temperature to power a certain reaction, and if the right ingredients aren't ready, then too bad. The temperature isn't going back up.

Stars, on the other hand, are stable thanks to a balance between the crush of gravity and the push of thermal pressure. Gravity and the other forces play off each other. For example, if the temperature is not high enough to support the star's own weight, then it will be crushed, contracting and heating until a new burst of nuclear reactions can keep gravity at bay. In a stable star, the temperature will adjust.

This balance requires a star with mass within a certain range. Too small, and the ball of gas will not be squeezed tightly enough by gravity to ignite nuclear reactions. Too large, and the star's over-excitable core will burn fuel very quickly, and is liable to blow off its outer layers of gas.[5] Our observations confirm these limits: nothing with fewer than 10^{56} particles shines, and no star with more than 3×10^{59} particles has been found. That's the *window*.

What happens, then, if we play with the strengths of the forces that govern stars?

Weakening the Pull of Gravity

Our first concern, upon weakening the pull of gravity[6], is whether stars form at all. Our Universe makes galaxies, stars and planets through the attractive pull of gravity; we will tell this story in the next chapter. For now we ask: in a universe with weaker gravity (and assuming stars form at all), what are stars like?

By weakening gravity, stars must be immense to be stable, which again raises the worry of how they would be formed in the first place. But there are more problems.

---

[5] Standard textbook lore says that stars that are more than about 100 times more massive than the Sun are unstable. They (supposedly) do not resist contraction or expansion, and so the slightest push leads to collapse or explosion. However, more sophisticated modelling has shown that so-called 'supermassive stars' can be possible: while they may be close to the instability limit in question, they do not cross it, and may be stabilized by relativistic effects and rotation. In fact, William Fowler (1966) showed that rotation could stabilize stars up to 100 million times heavier than the Sun. However, don't hold your breath for life's chances around such a star. They burn their nuclear fuel in about a million years. Thus, they would have to form on this timescale, drawing in more than a Sun's worth of material every year. When they exhaust their fuel, they either collapse into a black hole or produce the 'most energetic explosion in the universe', capable of completely emptying their host protogalaxy of its gas. We will treat the onset of short-lived, radiation-pressure dominated stars as a significant boundary for life, though more theoretical attention would clarify the situation.

[6] A technicality. What we are calling the 'strength of gravity' is also known as the gravitational coupling constant, α_g. It depends on the proton mass (m_p): $\alpha_g = G m_p^2 / \hbar c \approx 6 \times 10^{-39}$. Thus, changing the strength of gravity is equivalent to changing the mass of the proton, relative to the Planck mass. We use the proton mass because it is the most relevant mass for astrophysical objects; for example, the number of particles in a star is roughly $\alpha_g^{-3/2}$. We considered the effect on atoms and chemistry of changing the fundamental particle masses in the previous chapter. Here, when considering astrophysical objects such as stars, we will speak in terms of making gravity stronger or weaker because it's more intuitive. The mathematics is the same. Note well: given the range of the gravitational coupling constant that permits stable stars, this is (at most) *one* case of fine-tuning. It is not *two* cases, fine-tuning the strength of gravity and the mass of the proton.

These massive stars emit photons that are less energetic than those of their stronger gravity cousins. The photons emitted by our Sun are remarkably well matched in energy to molecular bonds. If the Sun put out gamma rays, molecules of life would be torn apart; conversely, if it emitted radio waves, the Sun would merely warm the Earth like a microwave oven. It would still provide low-entropy energy, of course, but not the kind that a molecular machine could use directly.

These feeble stars present another stumbling block to the formation of the complex elements needed for life: they are *boring*.

Stars in our Universe have two ways of transporting energy from core to surface. In *radiative* stars, photons carry most of the energy. In *convective* stars, the gas itself moves and churns, mixing the hot gas below with the colder gas above.

In our Universe, the line between radiative and convective stars runs right through the middle of our stellar window. Large stars tend to be more radiative, small stars tend to be more convective. Many stars have both convective and radiative zones.

Here's why this matters. In our Universe, just before a very massive star exhausts its fuel and explodes, it is layered like an onion. The inside parts are denser and hotter than the outer parts, so they burn fuel faster. When the core has burned to iron, which is as far as energy-producing nuclear fusion can go, layers of silicon, oxygen, neon, carbon, helium and hydrogen surround it. This is only possible because convection doesn't mix the layers. When the core explodes, the useful elements in the outer layers are sent out into the Universe. These *core-collapse supernovae* are responsible for about half of the carbon outside of stars.

In mid-sized stars, the core only burns as far as carbon and oxygen. A small amount of carbon is produced in the convective, churning outer layers of the star. As the star ages and runs out of fuel, its many contractions and pulsations throw off the outer layers. This ejected gas is called a planetary nebula, and accounts for the other half of carbon outside stars.

Small stars, however, are boring. They are convective through-out, and are thus able to effectively bring all the available fuel to the core to be burned. There are no layers. Very small stars, for example, burn all their hydrogen to helium, and then are not hot enough to burn further. They lock their matter away, becoming a *white dwarf* made of pure helium, a ball of matter supported against gravity and slowing cooling, but doing little else.

Let's return to our weak-gravity universe. Crucially, the radia-tive vs. convective line does not move in step with the mass window. In a weak-gravity universe, all stars will be convective. They will be larger versions of our Universe's small stars. They will burn through their fuel until they can't ignite any further reactions, and then stop.

The star could form layers if its core burns fuel faster than convection can mix the gas. Unfortunately, the star churns fast enough that this could only happen long after the star had burned carbon and oxygen into something else. Even if the star went supernova and spilled its material into interstellar space, it would be mainly heavy elements such as silicon and iron, rather than life-supporting oxygen and carbon.

On top of this, these stars may not go supernova at all. The neutrinos that expel the outer layers of massive stars in our Universe would be trapped in the core, unable to push effectively against the outside layers. They soon join the inevitable collapse into a black hole.[7]

Weak-gravity stars throw off a small amount of material in stellar winds, and may expel a little more when they exhaust their supplies of a particular element and switch fuel sources. However, for the most part, these massive stars die with a whimper and not a bang, with their essential elements locked away forever in their cores. The universe would remain a rather uninteresting sea of hydrogen

---

[7] The physics of core-collapse supernovae has proven difficult to unravel. It is believed that neutrinos are an important part of the explosion mechanism, and a simple timescale argument gives our conclusion here. See Carr and Rees (1979) for more details.

and helium, dotted with only mildly more interesting dead and dying stars.

The Curse of Stronger Gravity

While weakening gravity will lead to a rather boring cosmos, strengthening the force of gravity leads to a much more exciting universe. Well, exciting for a little while, but not necessarily suitable for life.

In stronger-gravity universes, our window for stars shifts downward to smaller masses. Stars are squeezed by the stronger gravity, and this intense pressure on the cores ensures the nuclear burning is accelerated, and the stars burn hotter and more rapidly, forging heavier and heavier elements. Even small balls of gas are able to ignite their cores and burst into stellar life.

These smaller stars burn hotter, emitting more UV radiation and even X-rays. Molecular-machine-powering radiation has been replaced with sterilizing gamma and X-ray photons. Life on a planet would need to huddle underground, letting the barren surface take the heat and trying to live off the residual warmth. As above, starlight fails to directly power chemical reactions.

With all this fast living, these stars die young, rapidly burning through their nuclear fuel. Stars would burn quickly through the universe's nuclear energy, emitting radiation that is lethal to life and dying in an even more energetic and dangerous supernova explosion. In our Universe, gravity is about 10^{40} times weaker than the strong force; if it were only 10^{30} times weaker, typical stars would burn out in a matter of *years*, not tens of billions of years.

Or so we thought, until Fred Adams of the University of Michigan took a closer look at stars in other universes. Fred discovered another condition on the stability of stars, a condition that narrows the stellar window as the strength of gravity increases. If gravity were 10^{35} instead of 10^{40} times weaker than the strong force, then the window would close completely. Stable stars would not be possible at all.

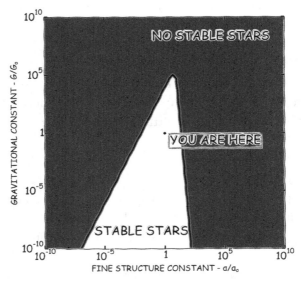

FIGURE 21 The stability of stars is a function of the strength of the electromagnetic (horizontal axis) and gravitational (vertical axis) forces. (Original figure appeared in Adams (2008).)

It appears, therefore, that if we want stable stars, stars able to steadily radiate the low-entropy radiation we need to drive life, we need gravity to be not just weak compared to the other forces, but extremely weak. Why this Universe obliges in terms of the strength of gravity is a mystery, and goes to the heart of the question of fine-tuning.

We've considered only the effect of changing the strength of gravity, but we've emphasized that all four fundamental forces govern stars. Electromagnetism is in action whenever positively charged nuclei repel each other, as well as in the scattering of light off electrons as it travels from the core to the surface. So, what if we *also* spun its dials?

Fred Adams's calculation also accounts for this; the results are shown in Figure 21. At first glance, things look OK. Clearly there are many combinations of the gravitational and electromagnetic forces that would result in no stable stars, but our Universe apparently sits in a triangle of stability.

In fact, that triangle has us cornered. We noted above that increasing the strength of gravity makes for unstable stars. In the figure, this means that the 'unstable stars' region hangs above the 'you are here' point. The extra freedom of changing the strength of electromagnetism – moving horizontally – actually makes things *worse*. Most moves to the left or right of our Universe make the unstable stars region *larger*.

Be careful! This figure is designed (quite rightly) to demonstrate the *shape* of the 'stable stars' region. It does not necessarily indicate the region's *size*. For example, the axes are *logarithmic*; instead of stepping 0, 1, 2, 3, . . ., they step . . ., 0.1, 1, 10, 100, . . . Furthermore, we have chosen the limits on the axes to isolate on the interesting bit. The plot continues beyond these limits in all directions.

Let's imagine changing the axes from logarithmic to linear. Let's extend the vertical axis – the strength of gravity – up to the strength of the strong force, as an estimate of what's possible. In this figure, the 'stable stars' region takes up less than one part in 10^{35} of the whole plot.

We've seen already how the strong and weak forces affect the stability of matter and the production of elements in the early Universe. With relatively small changes in these forces, nuclei are rendered unstable.

Even smaller changes *might* impact how stars burn. A small decrease in the strength of the strong force by about 8 per cent would render deuterium unstable. A proton can no longer stick to a neutron, and the *first* nuclear reaction in stars is in danger of falling apart. An increase of 12 per cent binds the diproton – a proton can stick to another proton.[8] This gives stars a short cut, an easy way to burn fuel. If the diproton were suddenly bound within the Sun, it would burn hydrogen at a phenomenal rate, exhausting its fuel in mere moments.

But there's a catch, or at least a complication. A universe in which the diproton was bound simply wouldn't make stars as hot and dense as

---

[8] Pochet et al. (1991)

the Sun. Remember that stars adjust to their reactions, so long as the star is stable. A stable diproton-burning star would have a cooler, less dense core. In practice, this means that smaller balls of gas would be able to ignite and sustain nuclear reactions. This isn't such a problem, and in fact would allow for stable stars within stronger gravity universes.[9]

More worrisome is the unbinding of deuterium. Stars would have to be massive enough to squeeze and heat the core until *three* protons managed to fuse directly into helium-3 (two protons, one neutron). In fact, another nudge to the strong force unbinds helium-3, so stars would require a *four-way* nuclear reaction. Igniting these reactions would require very large and thus very short-lived stars.

In summary, life requires stars to play a number of roles. They are where hydrogen and helium become all the elements of the periodic table. They are where your carbon, oxygen, iron, and more, were made. They are our source of useful energy, producing light able to power chemical reactions. They can do this for billions of years. And yet they are not perfectly stable – we need those elements, and so we need stars to die not with a whimper but with a bang. Stars are both the producer and distributor of the elements. And what's more, it is stars around which the debris of the explosion collects into planets.

These requirements translate into a precise set of conditions on the fundamental properties of the Universe. We need the strong force to power nuclear reactions and bind nuclei, electromagnetism to carry energy through and out of the star, and the weak force to change neutrons into protons and power supernovae. Above all, we need gravity to be weak: this, primarily, is why the Universe has large, long-lived, stable power sources.

If we think of a space of possibilities, our Universe sits on a delicate thread where stars are just right for life. It is possible, we suppose, that there are large chunks of this multi-dimensional *parameter space* that permit stable stars, but which we have missed.

---

[9] Luke is in the process of publishing a paper that uses models similar to those of Fred Adams to establish that stars in universes in which the diproton is stable can have similar properties to stars in our Universe. He will, inevitably, blog about it.

In this case, perhaps a dart thrown at random into our parameter space is likely to hit on successful stars after all. But even if these islands exist – and there is no evidence for them – we are still faced with the question of why this Universe, the one we inhabit, would sit on such a narrow thread when there is so much other potential parameter space we could inhabit. No matter how you look at it, the existence of stable stars appears to be a fine-tuned property of our Universe.

THE HOYLE RESONANCE

Before we leave the heart of stars, we need to discuss one of the most famous of fine-tuning cases: the Hoyle resonance. It is named after one of the great astrophysicists of the twentieth century, Fred Hoyle,[10] who is often remembered for some of his wackier thoughts on the history of the Universe and the origin of life. But he casts a long shadow over much of modern astrophysics, and especially the inner workings of stars. In the 1950s, as he was trying to work out the various ways that nuclear furnaces in stars forge heavier and heavier elements, he hit a snag. And that snag was the existence of carbon.

You may remember carbon from such objects as graphite pencils and diamonds. Scientific magazines breathlessly tell us that carbon's more unusual arrangements, such as buckyballs, nanotubes and graphene, have all manner of weird and wonderful (and hopefully technologically useful) properties.

Carbon plays an important role for life on Earth, as it forms the backbone for our molecular structures, such as proteins, carbohydrates, fats and nucleic acids. We are carbon-based life forms.

But why the focus on carbon? Could some other element play such a central role? Is there a possibility that we could have been neon- or vanadium-based life forms?

---

[10] We recommend Hoyle's autobiography (1994) and the biography by Simon Mitton (2011). Geraint had the pleasure of listening to Hoyle lecture on some of his more 'out there' cosmological ideas and, while the story was a little crazy, the spark of brilliance and broad Yorkshire accent still sticks in his mind.

As we've seen, the ability of elements to join together into molecules depends upon their distribution of electrons, especially those electrons orbiting furthest from the atomic nucleus. Some elements have filled their electron orbits, meaning that these *noble gases* are unwilling to join with other atoms to form molecules.

The outermost electron orbits in carbon are not full. In fact, carbon has four outer electrons willing to be shared with other atoms, and so it readily forms a myriad of different molecules. The variety and complexity of carbon molecules means that they receive special attention in modern science, with thick textbooks and research laboratories dedicated to *organic chemistry*, that is, carbon chemistry.

How many other elements could play the same role as carbon in providing the chemistry of life? The answer is not many.

Anyone who has watched the classic – and by 'classic' we mean 'made in the 1960s' – Star Trek episode 'The Devil in the Dark' will know that the fictional Horta (which looks suspiciously like a person crawling on the floor, covered by a blanket) is supposed to be a *silicon-based* life form. This aspect of science fiction is potentially science fact because silicon, while heavier than carbon, possesses a similar distribution of outer electrons, and so it too is chemically versatile. While the resultant molecular structures are not as conducive to supporting the chemistry of life, silicon-based life is a possibility.

Other than silicon, there is little hope for life based upon other elements. Boron and sulphur have been suggested, but these are very rare in the cosmos, and simply don't provide the long, linked, folding large molecules that life needs for its inner machinery and genetic blueprint. There are only 92 naturally occurring chemical elements, and carbon is by a wide margin the most suitable for life.

Now, we have reached a key question: where does all this carbon in the Universe come from? We've already seen that the intense fires in the earliest instances of the Universe forged helium out of hydrogen, but the fires diminished quickly and only traces of some heavier elements, namely beryllium and lithium, seeded the cosmos. In the immediate aftermath of the Big Bang, fewer than one nucleus in a hundred trillion is carbon!

We've also already seen that the furnaces in stars can burn for many millions and billions of years, and this burning is sustained through the fusion of smaller atomic nuclei into heavier elements. Stars create the carbon in so many of your molecules.

On the face of it, once a star has made helium, forming carbon should be simple; smash two helium-4 nuclei together to make beryllium-8 (4 protons, 4 neutrons), and then add another helium nucleus to make carbon-12 (6 protons, 6 neutrons). However, beryllium-8 is unstable. It will only hang around for 10^{-16} seconds before falling apart into two helium nuclei. We would need a star hot enough and dense enough that a third helium nucleus is likely to be on the scene within this short window of opportunity. Remembering that a helium-4 nucleus is also known as an alpha-particle, this reaction is known as the *triple-alpha process*.

The triple-alpha process was first proposed by Cornell physicist Ed Salpeter, and became the focus of Hoyle's research in the 1940s and '50s as he tried to unravel the inner workings of stars. Carbon, however, was proving to be a problem. As a star runs out of hydrogen fuel, it will no longer be able to resist the crushing force of gravity. It will start to contract and heat up, but it can't do this forever. Either it will ignite a new reaction, or be stabilized by *degeneracy pressure* (fermions resist being squeezed) and become a white dwarf or neutron star, or collapse into a black hole.

Hoyle's calculation was telling him that stars shouldn't be able to ignite helium. The necessary conditions were just too extreme – only when all of the hydrogen fuel is exhausted and the additional gravitational squeezing of the core drives the temperature over 100 million degrees Celsius does the rate of the creation of beryllium exceed the rate at which it disintegrates. Stars wouldn't reach this temperature, and so wouldn't make carbon.

Eventually, Hoyle hit upon an ingenious solution. He proposed that the carbon nucleus possesses a special quantum property called a *resonance*.

We know that an atomic nucleus is a ball of protons and neutrons, jiggling about together. If we inject energy into the nucleus, perhaps by firing in a high-energy photon, we can change the way that the protons and neutrons jiggle. The tiny world of the atomic nucleus is governed by the rules of quantum mechanics, and so the energy of a nucleus is *quantized*. That is, just like the electrons orbiting a nucleus, the protons and neutrons in a nucleus can only exist with very specific energy levels, but never at energy levels in between. Each of these distinct energy levels is a *resonance*.

Nuclei spend most of their time at their minimum energy level, known as the *ground state*. This is because a nucleus occupies a resonance or *excited state* for a very short period of time before losing this extra energy and changing to the ground state through the emission of high-energy gamma-ray photons.

The existence of a resonance has some interesting consequences for *making* atomic nuclei. Let's consider what happens when a beryllium nucleus and helium nucleus collide to form a carbon nucleus. Remember that these nuclei must approach each other at very high speeds to overcome the electromagnetic repulsion due to their positive charges. If they can overcome this, and get close enough for the strong force to get a grip, then these resonances come into play.

The newly formed carbon nucleus will contain buzzing protons and neutrons. The energy of this buzzing depends upon the energy of the collision between the beryllium and helium nuclei. If the colliding nuclei give carbon's pieces precisely the energy of the ground state, then the reaction can proceed very quickly as all the pieces can slot comfortably into place.

However, given the violence of the collision, the protons and neutrons in the carbon nucleus will almost certainly be buzzing about with much more energy than the ground state level. It will need to get rid of this energy. While there is a *chance* that this newly formed nucleus will shed the excess energy in the form of a gamma-ray photon and will settle down into the ground state, this is very unlikely. The collision is so energetic that the nucleus is too far from the ground state. More likely, the carbon nucleus will rip itself apart, returning to beryllium and helium.

However, if there were a resonance at just the right place in carbon, the combined energy of the beryllium and helium nuclei would result in a carbon nucleus in one of its excited states. The excited carbon nucleus knows how to handle the excess energy without simply falling apart. It is less likely to disintegrate, and more likely to decay to the ground state with the emission of a gamma-ray photon. Carbon formed, energy released ... success!

Thus, Hoyle reasoned that if there was a carbon resonance at an energy equivalent to the energy of a beryllium nucleus plus a helium nucleus plus some extra energy added due to the immense temperatures in the stellar core, then collisions between beryllium and helium would slip naturally into this excited carbon state. The presence of the resonance would boost the production of carbon in stars by a factor of 10 million.

By the 1950s, the nuclear properties of carbon, and many other elements, had been studied intensely in experiments. There was confusion about resonance levels in carbon, and certainly no strong evidence for the one Hoyle needed. Hoyle, however, knew exactly where to look, and badgered the experimentalists to try harder. In 1953, William Fowler found it, just where Hoyle said it would be.[11]

A brief but important aside. Hoyle's prediction is sometimes described as an 'anthropic prediction', in that he was supposedly spurred on by carbon's central role in life on Earth. Historically, this probably isn't true, and in any case life plays no part in the reasoning. Hoyle's inference – if no resonance, then no carbon – needs only the fact that the Universe contains significant amounts of carbon to conclude that the resonance must exist.

[11] History is never quite as simple as this. The interested reader should consult the detailed discussion in Kragh (2010).

The *existence* of the kind of resonance Hoyle needed is, in fact, fairly generic. It is called a *breathing mode*: the extra energy is used to expand and contract the nucleus without changing its shape. All nuclei have them. The trick, however, is to get the energy of the breathing right.

We can, therefore, ask whether the existence and properties of the carbon resonance are fine-tuned for life: if we varied the properties of the Universe, and in particular its fundamental parameters, would it still make carbon?

In 1989, Mario Livio and colleagues, of the Space Telescope Science Institute, simulated the life and death of stars with slightly different resonance energies. Relative to the ground state, a change of more than about 3 per cent begins to shut down carbon production. In fact, this happens for two reasons – either carbon isn't made at all, or else the star burns so efficiently that a carbon is made and then immediately burned up. The carbon nucleus can capture another helium nucleus, making oxygen.

While carbon is central to many of life's many molecules, the presence of oxygen in the Universe is also important. Oxygen plays a vital role in many biological molecules, and most famously in water and breathable air.

Our Universe has the rather remarkable ability to make both carbon and oxygen. Typically, stars in our Universe produce two oxygen atoms for each carbon atom. This occurs because of another fortuitously placed resonance. The breathing mode in oxygen is just below the level required to boost the rate at which carbon fuses with helium to become oxygen. This slows the reaction, leaving some carbon in the star, carbon that will eventually be ejected into the Universe to be used by life.

The rate of production of carbon and oxygen is quite sensitive to the strong nuclear force. If we nudge the strength of the strong force upwards by just 0.4 per cent, stars produce a wealth of carbon, but the route to oxygen is cut off. While we have the central element to support carbon-based life, the result is a universe in which there will be very little water.

Decreasing the strength of the strong force by a similar 0.4 per cent has the opposite effect: all carbon is rapidly transformed into oxygen, providing the universe with plenty of water, but leaving it devoid of carbon.

There is a worry, however, as the final stages of a star's life, during which these reactions happen, are a flurry of activity. Stars burn hydrogen at a leisurely rate, typically taking 10 billion years to exhaust their supply. But if large enough to ignite further reactions, the star will burn through helium in two million years, carbon in two thousand years, neon in a few months, and silicon in a few weeks. This increasingly rapid sequence, running out of one fuel and igniting another, produces complicated changes in the star, creating headaches for astronomers trying to understand the production of chemicals in the Universe.

As a result, a 0.4 per cent change in the strength of the strong force might not be as bad as we thought. It may take a slightly higher percentage to totally eradicate carbon or oxygen from the universe.

There is another fundamental property of matter that can affect nuclei, one that we have seen before: the masses of the quarks that make up the proton and the neutron. Steven Weinberg, the Nobel Prize-winning particle physicist, conjectured that the effect of playing with these masses would be small. The carbon nucleus, he reasoned, would act like a collection of three helium nuclei, and so it should not be surprising if there is a way to make them buzz around that looks like an unstable beryllium nucleus and a helium nucleus. This is just the resonance required.

Testing Weinberg's hunch, however, involves an extremely challenging and lengthy calculation. The precise energies of the resonances within a nucleus depend primarily upon the strength of the strong force, which dictates the binding of the protons and neutrons, as well as a smaller contribution from the electromagnetic repulsion of the positively charged protons. Calculating their combined influence in carbon, which has only six protons and six neutrons, might seem to be simple but, due to the complex nature of the strong nuclear force, it in fact requires supercomputer simulations. These calculational hurdles were overcome by physicists only in 2012, 60 years after Hoyle's initial prediction.

A team of researchers working at German and US institutes published the results of a calculation almost 10 million computational hours in the making. (This usually means a month of calculations on ten thousand linked computers, rather than one computer chugging along for a thousand years.)

Evgeny Epelbaum, Hermann Krebs, Timo Lähde, Dean Lee and Ulf-G. Meißner (2011) took up the challenge and considered what would happen to carbon and oxygen production if we mess around with the fundamental properties of matter. Contrary to Weinberg's hunch, the narrowness and location of the relevant energy levels translate into quite stringent limits on the masses of the quarks. A change of much more than a small percentage destroys a star's ability to create *both* carbon and oxygen.[12]

And remember from last chapter that because the quarks are already 'absurdly light', in the words of physicist Leonard Susskind (2005, p. 176), a range of mass that is a small percentage of their value *in our Universe* corresponds to a tiny fraction of their *possible range*. It is about one part in a million relative to the Higgs field, which gives them their mass. It is about one part in 10^{23} relative to the Planck mass!

So the fact that we are here typing these words, and you are there reading them, all constructed from molecules of carbon and oxygen, is only possible because the masses of the quarks and the strength of the forces lie within an outrageously narrow range!

THE INITIAL ENTROPY OF THE UNIVERSE

Before we close this chapter, there is one final but very important fact about entropy. As we have seen, by processing energy, life is

---

[12] This is a change of a small percentage of the *sum* of the masses of the light quarks (up and down). It is one fine-tuning, not two fine-tunings (that is, not one each for up and down.)

increasing the entropy in the Universe. In fact, all physical processes, from the cooling and collapsing of gas, to the burning of stars, to the violent explosion of supernovae, increase the entropy of the Universe.

However, entropy will not increase forever. Our Universe, so far as we can tell, will eventually contain no more free energy, and be effectively dead. This state of *maximum entropy* is the final destination of the Universe. It might take a very long time, but unless we are radically wrong in the way we understand the Universe, that's where we're heading.

What will the Universe look like as it ages, approaching its maximum entropy destination? We'll have to extrapolate a little, but the picture from the laws of physics as we know them is fairly clear. As stars burn, lighter elements are fused together. At the same time, very heavy elements decay through radioactivity to become lighter elements. There is a common goal: *iron*. Iron is the most strongly bound of the atomic nuclei, so it won't decay into lighter elements without an injection of energy, and it won't be built into heavier elements without other injections of energy. That is, it has no *free energy* to give.[13]

As the Universe ages, more and more matter is locked away inside iron, and so there is less and less energy to do things. Eventually, the material for running stars, the lighter elements such as hydrogen and helium, is effectively exhausted. The Universe will be left with fewer and fewer stars, feebly turning the remaining dregs of light elements into a faint glow, shining in an increasingly dark cosmos. One day, the last star will consume its last remaining morsel of fuel, and it too will go out.

Long before the last star is extinguished, the more massive stars will have died spectacularly, exploding violently as supernovae. These explosions are triggered when the core of a massive star fuses completely into iron and so ceases to give out radiation to resist gravity.

---

[13] In terms of physics as we understand it, the energy of an iron nucleus is permanently locked away. As we will see, new physics may offer an opportunity for this energy to be eventually released.

This is bad news for the star, as this radiation push is what supports the star against collapse. With the nuclear fires extinguished, the outer layers of the star fall inwards, crushing the core. A brief resurgence of nuclear activity blows the star into pieces. The core has completely collapsed, crushed into a black hole. And as the Universe ages, more and more of these corpses of giant stars litter the heavens.

Black holes are one-way portals, eating matter and energy but releasing nothing back into the Universe. Or so we thought, until Stephen Hawking convincingly argued in the 1970s that quantum effects near the black hole's point-of-no-return (the *event horizon*[14]) would result in particles being emitted from the black hole. This feeble radiation will slowly drain the black hole's energy, causing it to evaporate. The very low temperature of this radiation makes it almost impossible to imagine any useful process tapping a black hole. Black holes are the ultimate entropy machines.

In the very dim and distant future, the available free energy in the Universe will have diminished significantly, but there is still some to be had if you know where to look. The bodies of dead stars, those not massive enough to end their lives in supernovae, will litter the cosmos, and these will contain the remnants of chemical elements, mainly hydrogen and helium. To create a new star, we would need two of these objects to collide, and hope that their combined gravitational squeeze would ignite nuclear reactions. Such collisions, however, are extremely rare: expect the first one around 10^{22} years from now.[15] By this time, the new star will almost certainly shine alone in a dead Universe.

Alternatively, some future civilization, perhaps descended from the people, or cockroaches, alive today, could mine this remaining fuel. But they will have a limited time to do so!

---

[14] In both Disney's *The Black Hole* and 1997's *The Event Horizon*, black holes appear to be a portal to Hell. We contend that real black holes are much more interesting, though similarly hellish. However, black holes are not as complicated as the plot of *Interstellar*.

[15] Adams and Laughlin (1997).

On these long timescales, processes that don't show up in every-day life occupy centre stage. For example, we've seen how the weak force can transform particles. Left on its own, a neutron has a life of about 15 minutes, before it decays into a proton, an electron and a neutrino. But what about the proton, the lightest of the baryons – is it stable? The neutron can decay because it is slightly heavier than the proton, and so it is energetically possible for it to turn into a proton and a few other particles. The proton apparently does not have this option because it is the lightest member of its type of particles, those with three quarks (baryons), and so has no lighter cousin into which it can decay.

But what if, left to themselves for immense periods of time, protons could break the rules? What if they could turn into something other than their quark cousins? This would require a new force of nature, as none of the known forces allow for this. But – and here's why physicists take this idea seriously – such a force could be the missing link that unites the strong, weak and electromagnetic forces. This extremely feeble force would change the down quark in a proton into an electron, while flipping one of the up quarks into its antiparticle version, the anti-up quark. The electron, not being bound by the strong force, rapidly escapes, while the remaining up and anti-up quarks form a π^o, a meson known as a *pion*, a particle that rapidly decays into two photons. The proton is gone.

This force, affectionately known as X, has properties unlike any of the four fundamental forces. Figure 22 shows how X can turn a baryon (made of quarks) into a lepton, something that is never observed for the other forces. Physicists have searched for the telltale signs of proton decay, but have so far seen nothing. This could mean that this unknown force does not actually exist, but it could also mean that the force is very, very weak, and as a consequence the timescale for protons to decay is immensely long. The best estimates we currently have say that protons last at least 10^{32} years.

While 10^{32} years might seem like an awfully long time, it's nothing compared to the never-ending future awaiting the Universe.

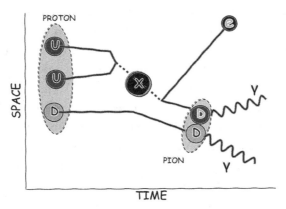

FIGURE 22 Feynman diagram of the potential decay of the proton, governed by an as yet undiscovered force, mediated by a new *X* boson.

We just have to wait, and eventually protons will decay, and matter will start to melt into a formless sea of low-energy radiation, neutrinos, and a few other trace particles. The expansion of the Universe will dilute this thinly spread energy even further. The Universe, facing a forever of eternal blackness, will have effectively died.

Our Universe would deserve quite a eulogy. In the (approximately!) 10^{100} years between its birth and the evaporation of its last black hole, its considerable reserves of free energy have driven all kinds of processes. Remember that these processes include everything that consumes energy, from making and drinking a cup of tea, to typing while sitting on a Sydney train, propelled along by electricity that gets its energy from condensed sunlight in the form of burnt fossilized timber, which itself was powered by the nuclear reactions at the heart of a collapsed gas cloud known as the Sun. Every step uses up a little of the Universe's vast store of free energy, winding down the Universe's ability to do useful things.

This rise of entropy over the lifetime of the Universe points back to a low-entropy beginning. In fact, our Universe must have been born with oodles of free energy to drive stars and civilization and such.

The seemingly special start to our Universe has bothered many great minds, so get yourself a strong cup of tea. We'll start at the beginning.

You'll remember that our Universe in its earliest moments is a smooth sea of hot plasma, with electrons, nuclei and radiation all bouncing around together. This featureless sea has an almost uniform temperature, and so you might think that it had little free energy. We said above that we need temperature differences to do useful things, right? However, we've neglected the elephant in the room: gravity.

Gravity tells mass to pull on other mass. The early Universe is thus set in motion, with small lumps of matter attracting other lumps. With a suitable set of machines, the energy of this motion could be extracted. When two lumps of mass have collapsed into each other, we've squeezed out all the energy available due to gravity.

So, the most useful arrangement of matter – from gravity's perspective – is spread out. A smooth matter distribution is, in fact, a low-entropy state, an ordered state of affairs!

As matter clumps together into stars, planets and galaxies, it increases its entropy and becomes more disordered. We will take a closer look at how the Universe got its galaxies and stars in the next chapter. But take a sip of tea and consider the entropy of the Universe for a moment. The apparently ordered Universe you see around you, with the structures vital for life to function and for you to exist, is actually the decayed remains of a more ordered state in the early Universe.

What would have happened if the Universe had not been born in its low-entropy state? What would such a universe look like?

Applying concepts like entropy to the entire universe is famously tricky, but a clever argument by the University of Oxford's Roger Penrose can help enormously.[16] Consider a universe that, unlike ours, doesn't expand forever but instead collapses into a 'Big Crunch'. It's like a Big Bang, but backwards. *Except* – and here is

---

[16] Penrose (1979).

the crucial point – the crunch will happen after a few billion years of entropy-increasing processes. Instead of a nice smooth Big Bang in reverse, the universe one second before the crunch will be full of lumps and bumps and black holes.

Now consider this lumpy, crunching universe, but expanding rather than contracting. *That's* what a high-entropy Big Bang would look like. The universe could have been born with matter locked away in black holes. This would rob the universe of much of the gravitational energy that ultimately powers life. All that would happen in these universes is that the black holes would evaporate over countless millennia, further filling the universe with low-energy, useless radiation. The prospects for sustaining life in such a cosmos are rather grim.

In fact, our Universe's entropy is freakishly lower than life requires. We don't live in a small island of order amidst a sea of disorder. Life at most needs a galaxy of order to create the right chemicals and stars; the rest of the universe can be high entropy, full of black holes and radiation, for all we care. And yet our Universe is orderly as far as the eye (and telescope) can see.

This order, from an entropy point of view, is an extravagance. Penrose has calculated the odds of the rest of the Universe being so orderly. Suppose, he begins, that life here needs an orderly region that is one tenth of the (linear) size of the Universe we can see. This is still a few billion light years, and so almost certainly an overestimate. Now, *if* the distant universe were arranged according to its entropy (so to speak), with high-entropy arrangements being much more likely than low entropy, what are the chances that it would be as orderly as our galactic neighbourhood? Penrose calculates: 1 chance in $10^{10^{123}}$.

That's quite a big number. If you wrote a zero on every single particle in the known universe, you still fall a long way short of writing down all the digits in that number. In fact, you would need roughly 10^{43} universes worth of particles just to write the number down. At this point, we suggest you take a minute to think about how big these numbers really are.

So, what do we do with Penrose's number? Somewhere in Penrose's calculation, an assumption has been made that results in an extraordinarily small probability for a true fact. Which assumption?

We'll revisit Penrose's calculation in the final chapter, but for now we must appreciate that the Universe we observe around us contains a truly astonishing amount of order. Even the order that we don't need for life here on Earth, that is seen only in the very distant Universe, is so abundant that it is difficult to believe that it is purely the result of chance.

You may have drained your teacup and be thinking that this is pointless speculation. Clearly, the Universe must have been born in a smooth Big Bang. Think back to Chapter 1. There is more to any scientific theory than the laws; equally important are the initial conditions. Laws say how physical things change, and initial conditions provide the starting point.

So, scientific theories do not predict their initial conditions. But surely the beginning of the Universe is the ultimate initial condition. How can Penrose then say that the beginning of the Universe, with its oodles and oodles of free energy, is *improbable*?

There are two reasons. Firstly, we aren't totally sure that the Big Bang *was* the beginning of the Universe. If someone comes up with an alternative theory of what happened in the early Universe that explains why it started in such a seemingly special state, then this would explain a great mystery. However, the odds are stacked against such an approach. Given the second law of thermodynamics – that entropy only increases – any explanation of the 'beginning' of our Universe in terms of a yet earlier state would need that state to be even more special!

Secondly, our conclusion that the beginning of the Universe is improbable depends on our understanding of physics. Penrose, for example, has argued that his calculation shows that there is a new law of nature that states that initial Big Bang singularities must be smooth. Perhaps we don't properly understand how to apply the concept of entropy to the very early Universe. Perhaps we're missing some important piece of physics.

These are exactly the options that fine-tuning for life leads us to consider. So far as we know, there are many ways that the Universe could have been born, many with much larger entropy at the start, and so with very little free energy. The wealth of free energy we have from our low-entropy start is quite surprising given what could have been. We should keep asking: why?

We will leave this question hanging at this point ... not that we know the answer, anyway. But there is more to say, and so we will revisit entropy when we talk about cosmological evolution in more detail in the next chapter. But it is important to realize that it is not only the fundamental properties of the physical laws, i.e. the masses of the particles and the strengths of the forces, that are finely balanced. Thanks to the ultimate dose of beginner's luck, our Universe was endowed with ample free energy for driving the processes that allowed us to be here. The question of fine-tuning applies not only to the cosmos we see around us, but also to whatever wound up the clockwork of the Universe in the first place.

5 The Universe Is Expanding

It is now time to go large and connect life with the Universe in its entirety. We will see how, from simple beginnings, the Universe expands and cools, forming galaxies, stars and planets. We'll see where our matter came from, consider dark matter and dark energy, and understand how cosmological theories reach back toward the very beginning.

So far, we've examined the question of what happens if we play around with the fundamental masses and forces. The results can be disastrous, as we can dramatically alter the make-up of the Universe, making it inhospitable for human life. We have also seen that, with a little more tweaking, we can make our Universe inhospitable to all life that we can conceive or imagine. But now it's time to zoom out by a few billion light years and consider the Universe as a whole.

INTRODUCING THE UNIVERSE

It can be easy to forget that we are living on the surface of a small, rocky planet, handily located at about 150 million kilometres from a typical star. This star, the Sun, is busily converting 600 million tonnes of hydrogen into helium in its core every second, eventually releasing the energy as sunlight which streaks through space and hits the surface of the Earth, warming it and providing the ultimate source of energy for life on our planet.

The Sun is not alone in the Universe, but lives with between 200 and 400 billion stars in a galaxy we know as the Milky Way.[1] The

---

[1] You may be shocked by how poorly we know the total number of stars in the Milky Way – how can we be uncertain by a factor of 2? The problem is that most stars are puny red dwarfs, which are difficult to see and so are difficult to count. If you want to think about this in terms of mammals, elephants are obvious and easy to count. Mice, on the other hand, may be much more numerous but getting a grip on their numbers is tricky due to their elusive nature.

Milky Way is immense, with light taking 100,000 years to cross from one side to the other, and is also richly structured; the Sun lives in a flattened disk of stars, with spiral arms lit up by recently created stars. At the centre of the Galaxy lives the bulge, a roughly spherical ball of older stars. Enveloping the Galaxy is a tenuous halo of stars, some of them the oldest we know, which must have been formed during the earliest epochs of the Universe.

We know that there is more to the Milky Way than meets the eye. As well as the atoms locked up in stars and planets, the Galaxy is permeated with gas clouds, mainly hydrogen and helium with traces of other elements. These elements are the remains of previous generations of stars, whose gas is blown into space as the stellar life reaches its conclusion, and it is an important part of the raw material that forms the next population of stars. While this gas is typically invisible to the naked eye, nearby bright stars can cause it to glow and become visible as the beautiful *planetary nebulae*. But even when un-illuminated, this gas can emit radio waves that are detected by telescope dishes around the planet. With radio-eyes, we would see the disk of the Milky Way glow brightly with the swirling clouds of gas that orbit around the Galactic centre.

While the gas in the Milky Way is difficult to see, we know there is another, more invisible component. The Galaxy is held together by a massive halo of *dark matter*, which is only detectable through its immense gravitational pull. The word *halo* is a poor description of where the dark matter lies, suggesting that it somehow sits like an angelic doughnut. It is really a big ball of, well, dark stuff, which penetrates right to the centre of the Galaxy, its density increasing rapidly.

It is important to remember that dark matter isn't something weird 'out there', but permeates the Solar System, and even the room in which you are sitting, the train on which you are riding or even the tree under which you are lounging. It is only because of the large mass of the Sun and Earth in your vicinity that this dark matter goes unnoticed. In fact, it is only the significant pull of the dark matter halo that keeps the Sun in its orbit about the centre of the Milky Way.

If all the dark matter instantaneously disappeared, the Sun would exit the Galaxy at high speed, destined to end its days in the darkness of intergalactic space.

Structure in the Universe does not stop with the Milky Way. Telescopes have revealed that the Galaxy is not alone in the Universe, but as far as we can see in every direction there are billions and billions of other galaxies. Some are very similar in size to the Milky Way, including our cosmic companion the Andromeda Galaxy, which is a mere two million light years away.

Like stars, galaxies come in a range of shapes and sizes, and our patch of the Universe is shared with almost a hundred smaller galaxies, known as dwarfs. The brightest and nearest of these, including the Large Magellanic Cloud, can be seen with our naked eyes, but most are puny things, with only a few tens and hundreds of millions of stars. This happy family of galaxies living together in our neighbourhood, and bound together by their mutual gravitational pull, is known as the *Local Group*.

In fact, most galaxies in the Universe live in such groups, sometimes with a handful of large galaxies, but always with lots and lots of dwarfs. But if galaxies live in bound groups, then there must be emptier patches of the Universe between them. These are quite prominent, and while we often see galaxy groups strung together like beads on a string, between them are these *cosmic voids* of relative emptiness.

Strings of galaxies and groups, separated by voids, can criss-cross each other, with the network of galaxies presenting a web-like structure. Where these crossings occur, we find the biggest, most massive bound structures in the Universe, namely galaxy *clusters*. These can be immense, containing tens of thousands of individual galaxies, all buzzing around in an extremely massive dark matter halo that holds them all together. And at the heart of clusters, we find very massive galaxies, given the innocuous title *cD*, which often contain many tens of trillions of stars.

The picture you should take away from this is that the Universe is richly structured on an immense range of scales, from individual planets and solar systems, up to the size of the cosmic web, covering billions of light years.

We are left with a rather imposing question, namely 'Where did all this structure come from?'

Maybe the Universe has always been this way? While that may be a comforting thought, over the last hundred or so years we've come to realize that the Universe is not static and unchanging, but is in fact dynamic and evolving, born in the hot, essentially formless soup of the Big Bang, and changing over cosmic time into the Universe we see around us today.

The Real Scandal of Cosmology

With its talk of expanding and curving space, dark energy and dark matter, cosmic radiation, and a 'Big Bang', one could be forgiven for thinking that *cosmology* – the study of the Universe as a whole – is a nightmare of complication. The trade secret of cosmology is ... get a little closer now ... closer ... *the Universe is simple*. Shockingly, mind-bogglingly simple. We can't believe we're getting away with this simple.

Now, this might surprise you a little. Anyone who has studied physics knows that it gets more complicated, more mathematical and more abstract as you go along. Algebra gives way to calculus. Calculus becomes vector analysis, and vectors transform into tensors. Physics seems to become an almost impenetrable sea of mathematics. And in the deepest depths of this sea sits quantum field theory, which is the basis of modern particle physics, and Einstein's General Theory of Relativity, the mathematical framework for cosmology. How can any of this be simple?

Let's explain.[2] Einstein's General Theory of Relativity relates the geometry of spacetime to the energy (of all forms, including matter) it contains. Roughly, you tell Einstein where the stuff is and he'll tell you how space and time intertwine and contort.

---

[2] For more, see Barnes (2014).

This *spacetime curvature* manifests itself as gravity. Gravity is not simply a force that pulls on particles, making them swerve. Rather, objects moving under gravity travel along locally straight lines, paths known as *geodesics*; energy distorts the very fabric of spacetime beneath them. Gravity doesn't turn the steering wheel; it banks the curve.

Peeking at general relativity's cogs and springs, which turn energy into spacetime geometry, we find a system of 10 coupled, non-linear, partial differential equations; for the mathematically uninclined, this is a bit like hearing your dentist say 'root canal', or your car mechanic say 'head gasket' – you might not know what those words mean, but be sure that pain is coming.

Applying these equations to the whole Universe, then, seems optimistic at best and a ticket to insanity at worst. So cosmologists in the 1920s and '30s – the early days of relativity – did what any good physicist does: they oversimplified. They assumed that on sufficiently large 'cosmological' scales, the Universe is the same everywhere (*homogeneity*) and looks the same in all directions (*isotropy*). This assumption was given the lofty title of 'the cosmological principle' but let's be frank – it's an optimistic guess, a toy model, a practice problem. The real universe surely can't be that simple. (The cosmologist Herbert Dingle (1953, p. 406) memorably warned his cosmological colleagues not to aggrandize a mere assumption: 'call a spade a spade, and not a perfect agricultural principle'.)

The result was the *Friedmann–Lemaître–Robertson–Walker (FLRW) model*; Friedmann–Lemaître–Robertson–Walker was not a quadruply barrelled member of the British aristocracy, but four giants of modern cosmology: Alexander Friedmann, Georges Lemaître, Howard P. Robertson and Arthur G. Walker. And it turns out that this simple picture is all we have ever needed. During the last hundred years of modern cosmology, various complications to the FLRW model have been investigated; none have improved on the original. The Universe is just about as simple as we could have hoped. Virtually every observation we have ever made of the Universe is explainable within this mathematical framework, a framework that can be fully written down in a page or two.

Again, you may be scratching your head, as cosmology appears to be an ever-changing field. It seems that hardly a week goes by without the media carrying a story about some new observation that has revolutionized our understanding of the Universe. (This week is no exception.) But while these observations tell us about the goings-on inside the Universe, the underlying mathematical framework is still that laid down by the cosmologists in the 1920s and '30s! This statement can shock those not in the field, but wander into any university library and pick up any book on General Relativity, flip to the pages on cosmology and take a quick look. The mathematical structure of cosmology is deceptively simple!

Curving and Expanding

The FLRW model describes two attributes of the spacetime of the Universe as a whole. The first is the *geometry* of space.[3] In the early 1800s, mathematician Nikolai Lobachevsky (the 'Copernicus of Mathematics', or for the less astronomically inclined, the 'Elvis Presley of Mathematics', famous for revolutionizing his field) showed that there is nothing unique about *Euclidean* or *flat* geometry, that is, the familiar geometry of high school where triangles have angles that add up to 180 degrees and parallel lines never meet. Mathematics tells of two other possible three-dimensional geometries that could describe the Universe, illustrated in the Figure 23. A universe as a whole could be positively curved, like a three-dimensional version of the surface of a sphere. Or else it could be negatively curved, somewhat resembling a saddle. Unfortunately, the two-dimensional version of this geometry is mathematically impossible to represent

---

[3] General relativity is a geometric theory, so it can tell us about the *geometry* of spacetime, but not the *topology*. That is, General Relativity can tell us how spacetime is locally curved, but not how it is globally connected.

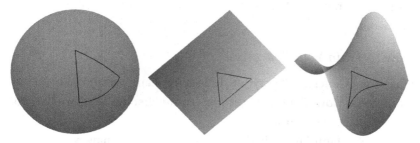

FIGURE 23 The three possible geometries of a homogeneous universe. On the left, a positively curved space. In the centre, a familiar flat space. On the right, a negatively curved space.

in three dimensions. If you can imagine a six-dimensional shape, let us know what it looks like.

The curvature of space is not merely abstract mathematics. It is measurable. If you find yourself in a spatially curved universe, and have a really big triangle and a lot of spare time, you will measure that its internal angles don't sum to 180 degrees. It matters not what the triangle is made of; the curvature is written into space itself.

Einstein's General Theory of Relativity tells us how the geometry of a universe depends on how much energy it contains. Overfill your universe and it will be positively curved – think of the aftermath of a big dinner. If underfilled, negatively curved. On the dividing line, flat Euclidean geometry holds, as seems to be the case in our Universe (on cosmological scales).

Secondly, the FLRW model describes the *scale* of space. Think of a model train – double the scale and all the parts double in size. In the case of a universe, it is not the contents but the scale of *space itself* that changes. On cosmological scales, beyond the reach of binding forces like gravity and electromagnetism, the distance between any two galaxies increases with time.

The Universe's space and time are no mere wooden floorboards, a static platform for the real actors. This stage curves, expands, and warps, responding to the energy that flows through it. (What a play that would be!)

How to Train Your Spacetime

Einstein's General Theory of Relativity can be summarized as *momentum flow shapes spacetime*. To understand that, we'll need to understand momentum.

We're familiar with the momentum of a runaway shopping trolley. More momentum means harder to stop. In classical physics, momentum is simply the product of mass and speed. Heavy things are harder to stop than light things, and fast things are harder to stop than slow things.[4]

The *flow* of momentum is familiar as *pressure*. Air pressure in a room measures how quickly the buzzing of air molecules deposits momentum onto the window. The more collisions (i.e. the denser the gas), the faster the collisions (i.e. the hotter the gas) and the heavier the gas particles, the more momentum flows and the larger the pressure exerted.

So momentum describes how particles move through space. When we move to a relativistic description of motion, we must add another component – the flow of momentum in the time direction. What is that?

It's easier than it sounds; it's just energy. Energy is the flow of momentum though time! Just as relativity combines space and time into one unified entity – spacetime – so it unites energy and momentum into one unified mathematical object, known as *4-momentum*.

We can return to our slogan: *(4-)momentum flow shapes spacetime*. All forms of energy and momentum gravitate. In particular, *pressure* gravitates. Positive pressure, such as the pressure of air inside a balloon, attracts just like mass.

---

[4] The notion that momentum is mass times velocity is ingrained in the minds of students entering university, but this is only true in classical mechanics, the world of large masses and slow velocities. It takes some effort to convince them that momentum is a more general concept and that light, even though it is massless, can carry it.

In short, Einstein's equations give stage directions to the different kinds of energy and momentum that could tread the boards. Now, we need a cast of characters.

The Playbill of the Universe

Compiling the Universe's *dramatis personæ* has not been easy. Many nights have been spent at telescopes taking a census of the sky. To measure light from the very early Universe, microwave detectors have been sent down to Antarctica, sent up in balloons, and even launched into space. Astronomers have used the motions of heavenly bodies to infer the presence of *dark matter*, while distant beacons of light betray the presence of the mysterious *dark energy*. Let's meet the entire cast.

When cosmologists talk about 'matter', they mean a form of energy that is (mostly) in the form of mass. Einstein's famous equation, $E = mc^2$, tells us how much energy is in a given amount of mass. Everything with mass is pulling on, and is being pulled by, everything else with mass. Matter hits the brakes on the expansion of space, and the more matter there is, the harder the brakes are hit.

'Radiation', by contrast, means a form of energy with no mass, with light being the most familiar example. The energy of a photon (particle of light) comes purely from its motion. The Universe's radiation, like its matter, pulls on its expansion, slowing it down. In fact, because photons have positive pressure as well as energy, radiation hits the brakes even harder than matter.

Astronomers have measured the amount of matter and radiation in the Universe. Galaxies shine with the combined light of hundreds of billions of stars, but most matter is not so spectacularly illuminated. Distant quasars, powered by the extreme amounts of energy released as matter makes its final plunge into a black hole, shine so brightly that we can see matter in silhouette. Between the quasar and Earth, we see the shadows of matter that has not collected into galaxies.

This tenuous, *intergalactic* matter is familiar – hydrogen and helium, mostly, with a smattering of heavier elements such as oxygen, carbon and magnesium. But we're still missing something. A cast member is unseen but active. Galaxies are rotating too fast to be held together by the matter we can see. Clumps of matter in the early Universe don't compress and rebound like they would if they were purely ordinary, proton–neutron–electron-based matter. When galaxies and clusters of galaxies collide, their gas compresses, heats, and glows – but most of their mass ghosts through, feeling the force of gravity but not the force of electromagnetism.

What is it? Answer that and collect your Nobel Prize. We know that it is matter, i.e. that its energy is mostly mass. We know that it doesn't interact via the electromagnetic force, which means that it doesn't emit or scatter light. This, for obvious reasons, is disappointing to the astronomer: all that stuff just sitting there, sending nothing to our telescopes!

We called it *dark matter*. Think of it as doesn't-send-us-any-light matter. Or, perhaps, we're-in-the-dark matter.

The matter – seen and unseen – and radiation in our Universe are not sufficient to halt the expansion born in the Big Bang. It appeared that the Universe would expand forever, always *slowing* but never *stopping*.

Still, something was missing.

Through the 1990s, evidence amassed that pointed beyond mere matter and radiation. The crucial clue came from observations of exploding stars, known as supernovae. Using Einstein's equations, we can calculate how bright these supernovae would appear in our telescopes, given energy's hold on the expansion of the Universe.

Einstein tells us that how we see something in the Universe depends on how the Universe has expanded in the time between the light leaving the source and arriving at our eye. How bright or faint something appears will depend, therefore, on the constitution of the Universe, as this is what controls the expansion.

With all the matter we see around us in the Universe, we expected its expansion to be decelerating. Thus, we expected distant

supernovae to be of a particular brightness. Two groups of astronomers, led by Brian Schmidt, Adam Riess and Saul Perlmutter, observed
that supernovae are actually *dimmer* than anticipated. In fact, they
were so dim that the Universe doesn't appear to be decelerating at all.
Einstein's equations tell us that the expansion must be accelerating.

Here's the conundrum. All the particles that you've ever seen,
and that astronomers have ever seen through their telescopes, and that
we've created in particle accelerators, cause the expansion of the
Universe to slow down. The expansion of the Universe is speeding
up. The stage is moving in ways that no *known* actor can cause.

What's going on out there?

We need another casting call: quirky character required for starring
role in accelerating spacetime. No time-wasters. We called this unfamiliar something *dark energy*. We're-even-more-in-the-dark energy.

What could make the expansion of the Universe accelerate?

Einstein's equations provide a potential solution, suggesting
that this acceleration is due to a strange property of spacetime itself
known as the *cosmological constant*. When Einstein derived his
famous equations, he was presented with a few forks in the road.
Being a good physicist, he took the simplest option. For example, he
could have considered twisty spacetimes (technically known as *tor-
sion*), but with no need to add this complication, he left it out.

At another point, Einstein could have added an extra term to his
equations that included the cosmological constant. Einstein's first concern with his equations was to reproduce the familiar pull of gravity in
the Solar System, and since the cosmological constant didn't help with
that, he left it out too. He took the simplest route first.

Einstein was a scientist of his time, and as there was no evidence
in astronomy for a changing and evolving universe, he, like others,
assumed that it was static and unchanging. However, he soon discovered that his equations would not allow for a static universe, because
with only the pull of gravity, the universe as a whole would not sit
still. So Einstein put his derivation in reverse, backed up to that last

fork in the road, and added a complication. With a cosmological constant, he could have a static universe.

In the following decade, the astronomers Vesto Slipher, Knut Lundmark, Georges Lemaître, Edwin Hubble and Milton Humason showed that the Universe was, in fact, expanding.[5] Because the Universe was not static, the cosmological constant was not required and so consigned to the 'dustbin of history', along with all the other historical clichés.

Einstein later supposedly called the introduction of the cosmological constant his 'biggest blunder', but it is important to understand why. What Einstein regretted was holding on to a static universe. Had he let that go, he could have predicted the expansion of the Universe years before it was observed. A missed opportunity! And in any case, Einstein's static universe didn't work. Arthur Eddington showed that it was like a pencil balanced on its end: the slightest push would have made it collapse or expand.

But while the cosmological constant isn't required to explain the *expansion* of the Universe, a positive cosmological constant acts like anti-gravity. So it could be just what we need to explain the *acceleration* of the expansion of the Universe.

Alternatively, an unusual form of energy could be causing the acceleration of the expansion of the Universe. In particular, we need a form of energy with negative pressure. This is not quite as weird as it sounds; negative pressure is simply *tension*. Inside a balloon, the motion of the molecules hitting the walls provides an outwards push, a positive pressure. The stretchiness of the balloon's rubber is an example of tension, opposing the outward push of pressure by trying to contract the balloon.

[5] The discovery of the expansion of the Universe is often attributed to Edwin Hubble alone. However, there has been a tremendous amount of historical research in recent years, clarifying who discovered what. Hubble remains central, of course, but many other astronomers contributed observational data and clues, as well as a community of theorists grappling with Einstein's new theory of gravity. See the collection of essays in Way and Hunter (2013), and Nussbaumer and Bieri's (2009) entertaining book, for more details.

Perversely, this negative pressure is a cosmic accelerator, driving the expansion faster and faster.[6] Whether the accelerated expansion of the Universe is due to a cosmological constant or a strange form of energy, cosmologists tend to bundle the two ideas into one: *dark energy*. It's a convenient shorthand for 'whatever makes the Universe's expansion accelerate'.

Our observations of distant supernovae reveal that 70 per cent of the energy in the Universe is dark energy. If only we knew what it was!

Since the initial discovery of dark energy, the evidence for its existence has firmed considerably, with its imprint being found on the distribution of the large-scale structure, the light of the cosmic microwave background (which we will meet in the next section) and even the motions of galaxies in the local universe.

Ladies and gentlemen, let us introduce to you, your Universe:

- 69% of our Universe is dark energy, whose negative pressure causes the expansion of the Universe to accelerate. We don't know what it is.
- 26% of our Universe is dark matter. We don't know what it is.
- 5% of our Universe is familiar, ordinary matter – the kind of which stars, planets, people and protons are made. Lots of this matter is diffusely scattered through the Universe, so we're not quite sure where it is.
- 0.3% of our Universe is in stars (that is, 6% of the ordinary matter), some of which we can actually see!

Similarly to the Standard Model of particle physics, we call this particular example of an FLRW model, which describes our Universe, the *Standard Cosmology*.

Because the various forms of energy dilute at different rates as the Universe expands, the top billing changes. In the very early Universe, radiation is the headline act. After about 100,000 years, radiation has diluted so much that matter takes over and continues with radiation's efforts to slow the expansion of the Universe.

---

[6] Just to be confusing: pressure pushes out, yet gravitationally pulls a universe in. Tension pulls in, yet makes the expansion of a universe accelerate outwards. We must distinguish the direct effects of a pressure *difference* from the *gravitational* effects of pressure.

Having waited patiently in the wings, dark energy began to direct the expansion just a few billion years ago, causing the Universe's transition to accelerating expansion. We find ourselves at a time when both dark components exert a similar pull and push on the expanding Universe.

Enter Photon, Stage Right

How do we know what the Universe is made of, and how much? Cosmology's Rosetta Stone is the cosmic microwave background (CMB). To understand why this light carries so much information about the Universe, we need to trace its path backwards in time. With the aid of Einstein's equations, we can hit rewind on cosmic history.

We'll be following the exploits of a typical batch of CMB photons coming from a particular direction on the sky. While there are hundreds of these photons in every cubic centimetre of space today, they go largely unnoticed, adding about 1 per cent of the static 'snow' on an out-of-tune television set.[7] Looking backwards, our batch of CMB photons entered Earth's atmosphere just 0.3 milliseconds ago.

Let's rewind our cosmic tape back 17 hours, and watch our batch of CMB photons reverse out of the Solar System and into the vast expanses of interstellar space. Indeed, it's called *space* for a reason. James Jeans, the great British astronomer, asked his readers to imagine three wasps in the air above Europe; that air would be more crowded with wasps than space is with stars. There is tenuous gas and dust between the stars, but only about one particle of matter for every thousand CMB photons, which is not enough to seriously affect our photon's journey. As we travel out through space and back through time, the light of the Sun fades, but our photons and their CMB siblings remain ubiquitous.

[7] Of course, modern televisions simply show a blue screen when untuned. But kids, ask your parents how much fun it was to tune in their favourite shows with the aid of a knob. Once you do, the saying 'don't touch that dial' should actually make sense.

A few thousand years ago, our photons were far above the Milky Way's disk of gas and stars. From there, you could see our Galaxy's majestic structure: its thin rotating disk of stars, white dwarfs, dust and gas, and its spiral arms where a new star is made every year (on average). You could also see our sister galaxy, Andromeda, about two million light years away. (Andromeda is approaching us at 300 kilometres per second. But don't panic. It won't be here for 3–4 billion years. And even then, remember the cloud of wasps. Would a wasp be worried about clunking heads with another wasp if it saw another such cloud approaching?)

Surrounding the Milky Way is a halo of dark matter. Keep that finger on rewind, and you'll see our batch of photons reverse out of our Galaxy halo about half a million years ago. They have now entered the space between the galaxies. They will spend 12.5 billion years in this vast desert, with just a few sprinklings of matter for company. During this stage of the flight, our photons can expect to pass through one galaxy, probably a smaller sibling to our Milky Way.[8] This will break the monotony for a few tens of thousands of years, before a return to intergalactic space.

The Evolving Universe

Let's consider how the Universe has changed during the flight of our CMB photons. Our Universe is expanding, so if we look into the past we would see that galaxies are closer together, and intergalactic gas more compressed. The Universe was denser in the past.

For the first 3 billion years of rewinding (that is, between now and 3 billion years ago), with each tick backwards, the Universe becomes more and more energetic. In the past, stars were being born more frequently, and more hungry black holes were being fed,

---

[8] How do we know this? Distant quasars are bright enough to see intervening gas in silhouette. Sufficiently dense regions of gas – known in the trade as damped Lyman-alpha absorption systems (DLAs) – are believed to trace galaxies and proto-galaxies that are assembling gas in preparation for their first burst of star formation. In 2000, Lisa Storrie-Lombardi and Art Wolfe showed that there is roughly one DLA along a given sightline in the last 12.5 billion years.

powering the active hearts of galaxies. Energetic radiation from stars and quasars filled the Universe, stripping electrons off intergalactic hydrogen; our Universe is mostly *plasma*.

But this activity peaks. Keep rewinding the tape back before 3 billion years ago, and each tick of the clock sees fewer stars and quasars being born. Around 12.8 billion years ago, there are too few stars to maintain the separation between electrons and protons, and so interstellar gas mostly consists of atoms of neutral hydrogen. But even more drastic changes are coming, as we look deeper into the past.

To our cosmic movie in reverse, the Universe appears to be *smoothing out*. Galaxies are not just producing fewer stars; they are being erased! Starlight returns to the gas clouds from which stars are born. We aren't exactly sure when, but around 13.5 billion years ago – just 0.2 billion years after the Big Bang – our photons are present at the (un)birth of the first galaxy.[9] The Universe enters the *dark ages*: there are no stars, but just a smooth soup of hydrogen atoms, 20 times denser than the Universe today, permeated by CMB photons. Our photons are still mostly spectators, and continue their journey back through the darkness.

The contraction of the Universe is making matter and radiation more dense, and hotter. Our photons, in particular, are becoming more energetic. We started with microwaves, with a wavelength of about a millimetre. With the Universe 20 times smaller than today, our photons are now infrared, and moving swiftly up the electromagnetic spectrum.

With the Universe now a near-uniform mix of matter and radiation, we can see that the CMB photons outnumber matter particles by

---

[9] You might think that galaxies are being erased because, with our finger on rewind, gravity's attraction would look like repulsion. Not the case. If you hit rewind on a video of the Earth orbiting around the Sun, it still looks like a small mass being attracted by a larger mass. Gravity is attractive in either time direction. Rather, the smoothness of the Universe in its earliest stages is not a result of its forces and laws, but is a special initial condition. As we saw in the last chapter, that's how our Universe started off.

about a billion to one. These particles of matter, however, are becoming a fog.

Our batch of photons passes harmlessly by neutral atoms in the Universe. But as the temperature increases as we look backwards in time, some of our bunch are energetic enough to knock electrons free from atoms. These free electrons are much more likely to bother a passing photon.

When the Universe is about 378,000 years old, and 1,000 times smaller than today, the fog of free electrons closes in. The Universe is so hot that electrons cannot stay bound to atoms. Somewhere in the very distant Universe, there is a free electron with our photon's name on it. The two collide. More than 13 billion years of travelling in a dead straight line is ended by a very sudden turn.

This electron is what we see in the CMB. Let's think about that a little more. To see anything is to see the *last scatterer*. Look around the room you are in; light comes in the window, scatters off a painting on the wall, and some of it travels into your eye. Your eyes and brain translate these photons into an image of the painting. Importantly, this is only possible because the air between the painting and your eye allows the light free passage.

Think about taking a photograph of the Sun. (Under no circumstances should you look directly at the Sun. How many times do you need to be reminded!?) Photons are created in the Sun's core, and take over a hundred thousand years to reach the surface, scattering all the way. Only when it has been released from the gas can the light stream to our camera. Each scatter erases any information about where the photon came from, and so our camera cannot see far into the Sun.

In contrast, the elusive neutrinos produced by the Sun's nuclear reactions in the core travel straight through the surrounding gas. Recently, scientists have managed to photograph the Sun in neutrinos, and see directly into its core. We think this is cool.

The Universe is like the Sun, but inside out. We see *out* to the Universe's dense, hot layers, from long ago, known as the *surface of last scattering*. We see hot, dense patches of matter from the Universe's early stages.

As a result, the CMB carries a wealth of information about the infant universe. Fluctuations in temperature and density in the patches of matter in the early Universe are imprinted in the patches of more and less energetic photons in the CMB sky.

Let's return to out intrepid batch of photons. They have been scattered, but not destroyed. They are bouncing around in an electron fog, but they were not born here. As we keep our finger on rewind, the fog continues to heat up, and our electrons become more and more energetic.

Knocking an electron off an atom is child's play compared with upsetting a nucleus. It's like the difference between swatting a fly off the elephant's back, and pushing the elephant over. At 378,000 years after the Big Bang, our photon horde can disturb the electrons buzzing around the outside of atoms. The nuclei, however, remain impervious.

Keep that finger on rewind, until just 20 minutes after the Big Bang. At this stage, the matter in the Universe is about 74 per cent hydrogen (single protons), 25 per cent helium (two protons and two neutrons) and a small amount of other nuclei.

The horde, which on Earth could barely raise static on a TV, is now unrecognizable. Around Earth, it takes a nuclear reactor or the crushing inferno at the centre of the Sun to fuse atomic nuclei together, or break them apart. In the first few minutes of the Universe, every photon is packing a nucleus-pulverizing punch. And they outnumber nuclei a billion to one.

The strong nuclear force, which has held helium nuclei together for the last 13.8 billion years of our backwards movie, succumbs to the rampaging photon army. Protons, buffeted on all sides, are sent crashing into nuclei. Within minutes the Universe is reduced to protons, neutrons and electrons. They cannot bind for more than a moment without being blasted apart.

The Universe is getting simpler. A second after the Big Bang, our marauding photon army has reduced matter to its building blocks.

What could finally stop such an army? Where did they come from?

Their final enemy is ... themselves.

In normal, Earthly circumstances, particles of light do not interact with each other. With flashlights, and even with lasers, you can cross the streams without fear. The photons pass each other by.

But two extremely energetic photons can do something amazing: they can create matter. This is matter–antimatter annihilation in reverse. When an electron meets an antielectron (a positron) the two can annihilate into two energetic photons. Now play it backwards. Two photons, if they have enough energy, can collide to create an electron and a positron. They need enough energy to make the mass-energy of an electron and a positron. The photons from your light bulb are short by a factor of tens of millions.

As the tape rewinds into the first second of our Universe, the cloud of photons explodes into a cloud of electrons and positrons. Because photons outnumber electrons by a billion to one today, each electron is suddenly joined by a billion siblings (and anti-siblings). The numbers of photons, electrons and positrons are roughly equal. Earlier still – before one hundred thousandth of a second – our photons are energetic enough to create billions of additional protons and antiprotons.

This is where our CMB photons were born. Our Universe began dense and hot, a seething, roiling morass of particles of every kind. Their collisions create and destroy each other – photons, protons, neutrons, electrons, neutrinos, their antiparticles, and more exotic, heavier particles.

For reasons that we do not understand, the very early Universe contains about a billion and one particles for every billion antiparticles. This may not seem like much, but we are all made from those billion-and-first particles. When matter and antimatter annihilate as the Universe cools, a remnant remains. From this, our Universe is made.

It is from this hot, expanding beginning that the Standard Cosmology gets its more popular name: the *Big Bang theory*. It is simply the notion that in the past the Universe was denser and hotter. That's it. We have seen in this section some of the consequences of that premise.

The Big Bang *theory* must be distinguished from the beginning of the Universe. In the FLRW model, the Universe begins at a point of infinite density and temperature, where the scale size of the Universe is exactly zero. Stephen Hawking and Roger Penrose first showed in detail how such a beginning is a *boundary* to spacetime.

However, this beginning is *not* an essential part of the Big Bang theory. The Big Bang theory is an extraordinarily successful scientific idea, explaining a broad range of observations. Whether the success of the theory means that the Big Bang points to the beginning of time itself is a different question entirely. The theory describes how the Universe was *raised*; how it was *born* – even *if* it was born – is another matter.

How the Universe Made Its Galaxies

We've told the story of the Universe's light, reaching from Earth back to the earliest moments of our Universe. There is another story to tell, not about light but about matter.

The cosmic microwave background gives us an accurate picture of the Universe 378,000 years after the Big Bang. The Universe contains a cooling gas of mostly hydrogen and helium. It is smooth and uniform, apart from tiny variations in density of around one part in 100,000. How does this expanding particle soup form the galaxies, stars and planets of the Universe today?

Don't underestimate this problem! We can see a time in the Universe's past where the largest deviation in density away from average was barely a thousandth of one per cent. Today, the air in this room is a billion billion billion times more dense than the average density of the Universe. An awful lot of matter needed to be collected from a large region of the Universe in order to make your high-density surroundings.

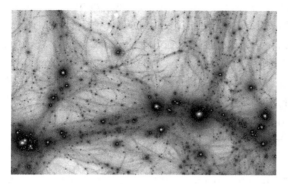

FIGURE 24 As gravity causes over-dense regions of the Universe to attract more mass and become more massive still, matter forms a cosmic web of sheets, filaments and knots. In particular, the knots are known as dark matter haloes, and are stabilized against further collapse by the random, buzzing motion of the dark matter particles. (Simulation by Pascal Elahi, Kevin Lam and Luke.)

Our story starts with gravity. Gravity is attractive – matter attracts other matter, and the more matter, the more attraction. The Universe is not egalitarian. The rich get richer, as larger lumps of matter attract smaller clumps and so grow larger still.

So the small lumps and bumps that we see in the early Universe, imprinted on the CMB, grow into larger lumps and bumps. Dark matter, being subject only to the force of gravity (as far as we can tell), has condensed and collapsed in this way for the last 13.8 billion years, forming the *cosmic web*. A typical blob of matter will collapse first along its shortest axis, creating a sheet. This sheet will continue to collapse, forming a thin filament. At the intersection of these filaments, matter will collect into dense knots. In these knots, the dark matter will be prevented from collapsing further by its motion: in the same way that the Solar System doesn't fall into the Sun because of the orbital motion of the planets, these dark matter knots maintain their size and density because of the random motions of the dark matter particles. These knots are known as *dark matter haloes* (Figure 24).

By contrast, ordinary (non-dark) matter cannot collapse freely in this way. Its thermal pressure, caused by the electromagnetic collisions of particles, can fight gravity by pushing back, refusing to be squashed. This is particularly effective when the Universe is young and hot. When the Universe becomes cool enough that electrons are bound into neutral atoms (about 378,000 years after the Big Bang), ordinary matter loses its fight, and goes along for the ride with its more abundant cousin by falling rapidly into dark matter haloes.

However, in our journey from lumps of matter that are one thousandth of one per cent more dense than average to lumps that are billions of billions of billions of times more dense, dark matter only gets us part of the way there. A typical dark matter halo is only about 200 times denser than the average density of the Universe. Actually, haloes tend to be about 200 times denser than the average density of the Universe *at the time they were formed*. Since the Universe was denser in the past, dark matter haloes that formed early in the Universe could be tens of thousands of times denser than today's average.

Even so, tens of thousands is a long way short of the factor of billions of billions of billions needed to create stars and planets. If not gravity, then what?

At first glance, it seems that the other forces of nature won't help at all. The two nuclear forces – strong and weak – have a short range, and only push and pull over distances comparable to the size of an atom. Electromagnetism is responsible for the forces of attraction between atoms in chemical compounds, but this only works inside molecules. In a gas, electromagnetic repulsion is responsible for individual molecules bouncing off each other whenever they come close. Atoms and molecules will attract each other in liquids and solids, but these require exactly the kinds of high-density conditions that we're trying to explain. How does matter collapse into galaxies?

In fact, electromagnetism is just the force we're looking for, but in the form of radiation. When two particles in a gas meet, they usually bounce off each other. But sometimes the collision jostles the electrons enough that a particle of light (a photon) is emitted. While the particles remain bound by gravity, the photon races off at the speed of light. It will soon exit the blob in which it was born, taking its energy with it.

This energy comes at the expense of the thermal energy of the gas, which is simply the random energy of motion (*kinetic energy*) of particles. If you could see the particles of air in this room, you would see them buzzing around at tremendous speeds (about 2,000 km/h!), colliding with each other and the walls and windows. In a blob of matter bound by gravity, these random motions support the gas against the attractive force of gravity, just as for dark matter.

So, ordinary matter leaks energy. We don't notice it in everyday life, but ordinary matter, having fallen into dark matter haloes, leaks away its ability to fight gravity and begins to collapse. (Particles of dark matter, meanwhile, don't collide with each other and so maintain their random motion.) Under the right conditions, this collapse will run away: particles collide, energy (from random particle motion) leaks away, gravity can crush the blob a bit more, the matter gets denser, the particles collide more often, more energy leaks, more collapse, and so on. Inside and around dark matter haloes, matter plunges towards the centre, to higher and higher densities. We're on our way to making a galaxy!

Where will it end? Why doesn't the matter simply collapse into a black hole? The reason is another type of motion: rotation.

The initial blob of collapsing matter will contain a small amount of rotational motion: there will be some axis that we can draw through the blob such that, if we add up all the matter going clockwise and all the matter going anticlockwise, the two will not exactly cancel. While the blob is large and diffuse, this hardly matters. But as the blob starts to collapse, rotation starts to become significant.

Consider a twirling ice-skater. With his arms outstretched, he rotates slowly. But as he pulls his arms in, he picks up speed, rotating

FIGURE 25 Three spiral galaxies. When a swirling, collapsing cloud of gas settles into a dark matter halo, it forms a disk galaxy. Stars form a thin disk that is stabilized against further collapse by its rotation. (A small patch of sky from the Pan-Andromeda Archaeological Survey, courtesy of Rodrigo Ibata of the Strasbourg Observatory.)

faster and faster. The effort expended in pulling his arms into a tighter orbit results in a faster spin.

As our proto-galaxy collapses, it begins to spin faster. The random motion of the gas is supplemented by an ordered rotation around a central axis. This provides a new stability, one that will remain even if the gas continues to cool by leaking radiation. Furthermore, any cloud of gas that wanders randomly through the rotating gas will collide with it, leak its energy and join the rotation. This includes gas that is rotating about a different axis to the bulk of the galaxy. The result is a rotating *disk* of matter, such as we see in the galaxies around us, which is much smaller than the dark matter halo in which it formed (Figure 25).

It is here, in the gaseous disks of newly formed galaxies, that stars and planets can form. But our galaxy's story is not yet finished. It remains a part of the cosmic web. In the Universe, large haloes grow by feeding on and merging with smaller haloes, known as *hierarchical* formation. Within this tangle of condensing matter, galaxies are not

FIGURE 26 The Coma Galaxy Cluster is about 300 million light years away, and consists of thousands of galaxies orbiting each other. The entire cluster weighs about as much as a hundred Milky Way galaxies. (Source: Mike Irwin, University of Cambridge; used with permission.)

evenly spread through the Universe, but will collect into groups, clusters and super-clusters. (Figure 26). A good example is our own cosmic neighbourhood, imaginatively titled the *Local Group*. About a hundred smaller galaxies join the Milky Way and its similarly sized sister, Andromeda, and new dwarf galaxies are being discovered all the time.

With so many galaxies buzzing around, it is no surprise that they occasionally collide. Such collisions take billions of years, so our astronomical observations provide us with a snapshot of these cosmic train wrecks. Through a series of snapshots, we can piece together a *merger sequence*, as shown in Figure 27. The disk of stars and gas is warped by the gravitational pull of the other galaxy. The gas collides, compresses and creates a burst of new stars. So-called *tidal streams* of gas and stars are torn from the outskirts of each galaxy, forming characteristic pointers back to the impact site.

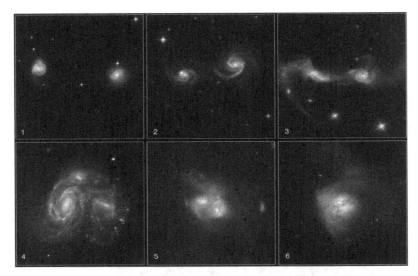

FIGURE 27 A selection of interacting galaxies, chosen to illustrate a merger sequence. As the galaxies approach and collide, their mutual gravity draws streams of gas and stars into thin *tidal streams*. In the collision, their gas compresses and collapses further until the two galaxies become one. (Credit: NASA, ESA, the Hubble Heritage Team (STScI/AURA)-ESA/Hubble Collaboration and A. Evans (University of Virginia, Charlottesville/NRAO/Stony Brook University), K. Noll (STScI), and J. Westphal (Caltech); used with permission.)

While the sky provides snapshots, a simulation can reveal all the gory details of the crash, using computers programmed to solve the laws of physics. These simulations guide our understanding of galaxy mergers in progress in the real Universe, revealing the violence of galaxy metamorphoses (Figure 28). Depending on the orientation of the colliding disks, the final result could be a new, larger rotating disk, or a rounder ball of gas and stars.

The Universe is littered with the results of galaxy mergers, the most conspicuous being the abundant population of so-called *elliptical* galaxies (Figure 29). You may not have seen a galaxy like this before, and for good reason: they're not as pretty as their disky cousins.

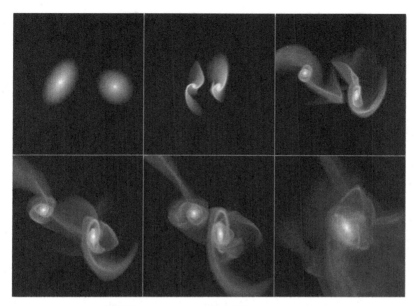

FIGURE 28 Using computer codes that incorporate gravity, stars, gas pressure and more, we can simulate what a collision between two galaxies would look like. Compare this image with the previous one, noting especially the tidal tails and the 'train wreck' aftermath. (Image courtesy of John Dubinski; used with permission.) For mesmerizing animations of galaxy mergers set to music, visit www.cita.utoronto.ca/~dubinski/galaxydynamics/.

They are, nevertheless, about as common as disk galaxies, and usually heavier, with older stars and lacking the material to make new stars.[10]

While galaxy mergers play an important role in turning disks into ellipticals, there are more actors in this play. If a galactic disk is agitated, such as in a close encounter, then instead of settling back into a neat ordered rotation, it may find that its innermost parts warp

[10] Astronomers love to classify things, to put things in boxes, although some astronomers' obsession with classification is more akin to stamp collecting than science. And so it is with galaxies, with ellipticals put in a box called *early-type*, while their spirals as known as *late-type*, based upon where they were drawn on what became known as the 'tuning-fork' diagram. Some think that the name refers to some sort of evolutionary sequence, with ellipticals becoming spirals, but the evolution actually appears to go the other way, with the collisions of spirals creating ellipticals.

FIGURE 29 Two elliptical galaxies. This type of galaxy is an elongated ball of stars, with very little gas and less ordered rotation than a disk galaxy. (A small patch of sky from the Pan-Andromeda Archaeological Survey, courtesy of Rodrigo Ibata of the Strasbourg Observatory.)

and wobble and widen into a galactic *bulge*. Galaxies we can see in the very distant (and therefore, very early) Universe seem to be much more lumpy than the neat, ordered disks we see today. This *disk instability* has an important relationship to the formation and migration of stars, though astronomers are still working out the details. It seems that galactic disks typically first form a bulge, then their stellar population becomes on average much older, and then star formation shuts off. Or, in the words of Australian astronomer Ned Taylor: first you get fat, then you get old, and then you die.

The occupants of galaxies, too, play their role. When massive stars exhaust their nuclear fuel, they explode in supernovae. If enough supernovae go off together, their combined explosive effect can drive gas out of galaxies and back into intergalactic space. We see evidence of this process in the form of galactic winds: some galaxies feature highly energetic particles of matter streaming off their disks. Also, at the centre of most galaxies is a black hole, which can be heavier than a billion Suns. Matter falling into the extreme gravity of a black hole forms a rapidly rotating, extremely energetic disk. This disk will be millions of times smaller than the disk of gas and stars in the galaxy, but will produce copious amounts of radiation. Like supernovae, this radiation can heat and expel gas from galaxies. The understanding of such *feedback* processes is the focus of much galaxy formation research today.

Within galaxies, gas is dense enough to form stars and planets. Again, gravity is the driver: lumps of matter collapse under their own weight. In stars, the collapse of gravity is eventually halted by the ignition of nuclear reactions in their cores. For planets, the ordinary kicking-rocks-hurts rigidity of matter is enough to prevent gravity from crushing a ball of rock into a black hole.

Returning to the big picture, the Big Bang theory is doing a marvellous job of explaining the Universe we see through our telescopes. Alternative cosmological ideas abound, as a Google search will reveal. But these must run the gauntlet of the entirety of the data, and, to date, none come close to explaining the Universe in the detail provided by the Big Bang theory. Puzzles remain, so we will continue to observe and simulate, think and theorize. Nevertheless, some features of the Universe, as we understand it, strike cosmologists as unusual to say the least.

COSMOLOGICAL PROBLEMS

The Problem with Dark Energy

If you've read carefully between the lines, and followed our cryptic clues, you should have worked out what dark energy is by now.

Just kidding. We have no idea.

Actually, we know a few basic things about dark energy. We know that it makes up 69 per cent of the energy of the Universe. We know that it causes the expansion of the Universe to accelerate. And from this, we know *something* about its character, and in particular how it enters our cosmological equations. So we can ask: what would the Universe be like if dark energy, as we understand it, was different? Let's start by simply changing the amount of dark energy.

To cut a long story short: too much dark energy shuts down galaxy formation. For as long as the braking effect of matter controls the expansion of the Universe, dark matter haloes will collapse and merge and collect more matter from their surroundings. But when matter has spread itself so thinly that dark energy takes over, the accelerated expansion of the Universe dilutes its matter faster and faster. Galaxies and their haloes are driven apart – too far to merge, too far to collect more matter from intergalactic space. With their supplies cut off, galaxies use what gas they have to make stars, and then shut down as stars burn out and die.

In fact, this era lies in the future of our Universe, when its dark energy comes to dominate. But in other universes, with a little more dark energy, this era comes earlier, shutting down galaxy formation before it starts. The accelerated expansion of the Universe will push matter out of gravity's reach. Matter will not collapse.

In these universes, any useful free energy from forming stars, planets and people is so rapidly diluted as to be useless. These cold and lonely universes are doomed to be lifeless.

There is another possibility to consider: dark energy can be *negative*. Negative energy isn't the kind of property possessed by matter or radiation. But we know so little about dark energy that it remains a possibility. In particular, Einstein's cosmological constant could have been negative.

Negative dark energy means that a universe is heading for a big crunch. Like matter, negative dark energy causes the expansion of the universe to decelerate. The difference is that the effect of dark energy doesn't dilute with the expansion of the universe. The universe cannot escape its pull, and expansion must eventually become contraction. And then comes the crunch.

A small amount of negative dark energy will not significantly affect life. Such a universe will live long and prosper, making plenty of galaxies and stars before its crunchy end. But as we increase the amount of negative dark energy, the end comes sooner. Too much, and the end comes before galaxies, stars, planets and life.

So, the energy content of our Universe appears to be walking a fine line. If our Universe had had a differing mixture of dark energy and matter to the one we observe, its evolutionary history would have been dramatically different. With dark energy, a universe is rather easy to ruin.

The Real Problem with Dark Energy

Now we come to a dirty secret of modern physics, an embarrassment that is introduced to students in a popular textbook as,

> ... arguably the most severe theoretical problem in high-energy physics today, as measured by both the difference between observations and theoretical predictions, and by the lack of convincing theoretical ideas which address it.
>
> (Burgess and Moore, 2007, p. 439)

(Thinking of a PhD in physics? We need help!)

The problem is with the identity of dark energy. While there are lots of possible culprits, one of the main suspects is something called the *quantum vacuum*.

Our best theory of how matter behaves is known as quantum field theory, which we met in Chapters 2 and 3. Let's pull that idea apart. A field in physics is a physical entity that exists at *every* point in space and time. In contrast, a particle exists at a *particular* point at any given time.

The unique feature of *quantum* fields is that certain configurations of the field describe the particles around us. Wiggle the field in a certain way, and it will move and interact like a particle. Or two

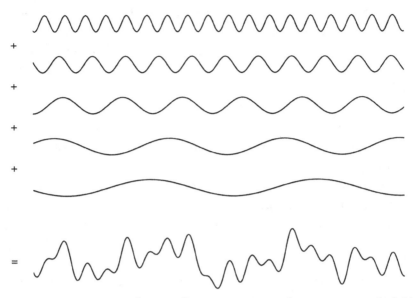

FIGURE 30 Mathematically, we can represent the vacuum state of a field as a collection of waves, each with its own characteristic energy, which add up to give the total energy of the vacuum.

particles. Or loads of particles. Or no particles. To quantum field theory, an electron is nothing but a wiggle in the electron field.

In fact, all of the matter in the Universe consists of wiggles in these quantum fields: the electron field, the quark fields, the photon field and more. Their wiggling and mutual interaction underlies the motion and action of everyday life.

So, if we empty some space of all the particles – a state known as the *quantum vacuum* – the field remains. The crucial question for us is: how much energy remains in the field? Remember that the field is still there, and can still wiggle, when no particles are present.

In Einstein's theory of gravity, this *vacuum energy* acts like dark energy, causing the expansion of the universe to accelerate. If we could calculate the energy, we could predict the amount of dark energy in the universe. This could be yet another triumph for quantum field theory!

Each field contributes to the vacuum energy. Moreover, each field's vacuum state can be represented as a succession of waves. Figure 30 shows how waves add together. Each contribution has a characteristic energy.

If we naively add up all such waves, we get the answer *infinity*. That's no good. Infinity usually means we've made an unwise assumption, and in this case it's not difficult to identify. We've assumed that quantum field theory remains an accurate description of matter at infinitely high energies. Clearly, this far outstrips our experimental evidence.

In fact, it's a suspicious assumption on theoretical grounds also. There is a natural energy scale, known as the *Planck energy*, at which we believe that gravity can no longer be ignored. Quantum field theories do not include gravity, and a simple calculation suggests that a quantum particle with Planck energy would become its own black hole. That might not actually happen, but we certainly can't trust any theory that ignores that!

Including only energies in the vacuum up to the Planck scale results in a finite prediction for the energy in the quantum vacuum, which is better than infinity. But how does it compare to the value required to explain our observations of the accelerated expansion?

The prediction is 10^{120} times larger than the observed value. Just pause to soak in that number's colossal magnitude. This answer is staggeringly wrong.

Let's think about you, an average human being. How many protons are in your body? A quick calculation reveals this to be around 10^{28}, a pretty big number, but virtually nothing compared to 10^{120}. The number of protons in all the people on Earth is about 10^{38}, and the number of protons in the Solar System is roughly 10^{57}; the numbers are getting larger, but are still not quite there.

Let's go a little larger and ask how many protons are there in the observable Universe? We can count the number of stars and galaxies, and we can estimate that there must be around 10^{80} protons. Surely, you must be thinking, we must be almost there, but in fact we are not even close.

If there was an individual observable universe for every proton in the body of every human on Earth, then the sum of all of the protons in all of this multitude of universes would be around 10^{118}, still a factor of 100 times smaller than the discrepancy between our predicted energy density in the vacuum and that we actually observe. Now you can see why this is rather embarrassing.

But wait! The Planck energy is a very long way above any energy we've produced in an experiment – about a thousand trillion times higher, in fact. We should be more conservative about how much we trust our theories.

Suppose we pick an energy that has been comprehensively probed by particle physics experiments, and below which quantum field theory has performed with distinction. The discrepancy is reduced by a factor of 10^{60} – success! Except, even with our new calculation, you still disagree with nature by a factor of 10^{60}. It's a start, but keep the cork in the champagne bottle.

Maybe there is a mechanism at work here, a mechanism that we clearly don't yet understand, which trims the energy in the quantum vacuum; so, while it is intrinsically very large, the value we observe, the value that influences the expansion of the Universe, appears to be much, much smaller. But this would have to be a very precise razor, trimming off 10^{120} but leaving the apparently tiny amount that we observe. It would seem natural for such a process to remove the presence of the quantum vacuum completely, rather than leaving the tiny residue we see about us.

But what if this mechanism for supressing the influence of the cosmic vacuum energy was not so efficient, removing the effect of 10^{119} rather than 10^{120}, so there would be ten times the vacuum energy density we actually measure? Remember, such vacuum energy accelerates the expansion faster and faster, emptying out the Universe, cutting off the possibility of stars, planets and eventually people.

If the Universe faced the raw energy of the vacuum, it would be empty and dead.

This is the *cosmological constant problem*, though the name is a slight misnomer. The *effective* cosmological constant is the sum of Einstein's cosmological constant and all the forms of energy in a universe that behave like a cosmological constant. Vacuum energy is such a form of energy. The accelerated expansion of our Universe is due to the effective cosmological constant. So the problem is this: why is the effective cosmological constant so much smaller than each of the vacuum energies?

The cosmological constant problem is not as straightforward as 'prediction gone wrong'. Vacuum energies can be positive or negative, so summing all the different *energies* could miraculously make them cancel each other out to 120 decimal places. But still, it's not very likely.

As an example of fine-tuning, the cosmological constant problem is a near-perfect storm. This is important so, in some detail, here's why.

1. It's actually several problems. Each quantum field – electron, quark, photon, neutrino, etc. – adds a ludicrously large contribution to the vacuum energy of the universe.
2. General Relativity won't help. Einstein's theory links energy and momentum to spacetime geometry. It does not dictate what energy and momentum exists in the Universe. Universes that are no good for life are perfectly fine by the principles of General Relativity.
3. Particle physics probably won't help. Particle physics processes – those described by quantum field theory – depend only on energy *differences*. We can change the absolute values of all the energies in particle physics and all the interactions remain the same. It is only gravity that cares about absolute energies. Thus, particle physics is largely blind to its effect on cosmology, and thereby life.
4. It isn't just a problem at the Planck scale, so quantum gravity won't necessarily help. As noted above, we don't need to trust quantum field theory all the way up to the Planck energy in order to see the cosmological constant problem. It is entrenched firmly within well-understood, well-tested physics.

5. Alternative forms of dark energy have exactly the same problem. Dark energy – whatever is making the expansion of the Universe accelerate – might not be vacuum energy. But alternative forms of dark energy usually posit some other kind of field, and so the problem of the vacuum energy of the field returns, unchanged and unsolved.

6. We can't aim for zero. Before the accelerated expansion of the Universe was discovered, it was thought that some principle or symmetry would set the cosmological constant to zero. Even this was a speculative hope, and it has since evaporated.

7. The quantum vacuum has observable consequences, and so cannot be dismissed as mere fiction. In particular, an electron in an atom feels the influence of the surrounding quantum vacuum.[11] Our theory works beautifully for electrons and atoms. Why doesn't cosmic expansion feel the full influence of the quantum vacuum?

8. The (effective) cosmological constant is clearly fine-tuned. It's just about the best fine-tuning case around. There is no simpler way to make a universe lifeless than to make it devoid of any structure whatsoever. Make the cosmological constant just a few orders of magnitude larger and the universe will be a thin, uniform hydrogen and helium soup, a diffuse gas where the occasional particle collision is all that ever happens. Particles spend their lives alone, drifting through emptying space, not seeing another particle for trillions of years and even then, just glancing off and returning to the void.

And *that's* why we need help!

Does This Matter Make Me Look Flat?

As we noted above, the contents of the universe determine its geometry. In particular, a *critical energy density* separates the over-dense, positively curved universes from the under-dense, negatively curved universes. Right on the dividing line, universes are flat, with Euclid's familiar geometry.

Our Universe, it turns out, is so close to the critical energy density that we don't know whether we are exactly flat or very slightly curved. Our measurements are accurate to about one per cent, but we

---

[11] For the experts, the Lamb shift (Polchinski, 2006).

are so close to the dividing line that this isn't enough to determine what side of the line we are on.

This makes cosmologists slightly uncomfortable, for the reasons explained in Chapter 1. A physical theory with a free parameter is really a collection of theories, one for each value of the parameter. It will look suspicious, then, if we need to fine-tune that parameter. If only a small fraction of the collection of theories explains the data, then the door is open for an alternative theory.

Returning to our Universe, having a density that is within one per cent or so of critical isn't too suspicious. The real problem is that this is the Universe's density *today*. Let's follow the Universe back to discover its initial conditions. When we do, fine-tuning suspicions are aroused.

While the expansion of a universe decelerates, curved universes become more curved. *Exactly* flat universes remain flat, but the slightest curvature is quickly amplified. Because (according to the Standard Cosmology) our Universe was decelerating for the first few billion years of its existence, its initial conditions must be preposterously fine-tuned for it to appear flat today. Suppose we wind the tape of the Universe back to when the first elements are formed, a few minutes after the Big Bang. Then, in order for our Universe to be flat to within one per cent today, it must have been flat to within one part in a thousand trillion (1 followed by 15 zeros). Suspicious!

Winding the clock further back in time only makes it worse. The furthest we can meaningfully wind the clock back is the so-called *Planck time*, before which we would need a theory of quantum gravity to predict what's going on. (We don't have one. Or at least, not one that is well understood and well tested.) The Planck time is equal to 10^{-44} seconds. At this time, the fine-tuning is one part in 10^{55}. Even more suspicious!

This is known as the *flatness problem*. It is a problem for cosmology. It is a problem for life, too.

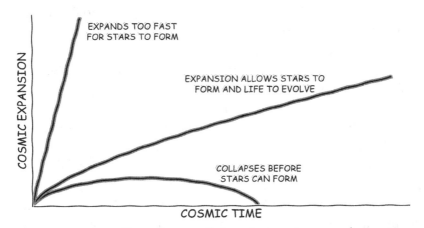

FIGURE 31 Depending upon their contents, universes can have quite different expansions, with significant influence on life therein.

To understand this, let's go back to those initial conditions and play with them (Figure 31). We'll start by increasing the amount of matter in the Universe, and so increasing the braking effect it has on the cosmic expansion. We don't have to increase it too much before it really slams on the brakes. The expansion is soon brought to a halt, and comes crashing back down. The universe is destroyed in a big crunch.

What are the prospects for life in these James Dean and Kurt Cobain-like universes, living fast and dying young? As we increase the initial density of matter in these universes, they experience a long lifetime of many billions of years, and conditions will be similar to the Universe we presently inhabit. But as we continue to wind up the density, the lifetime of the universe becomes shorter and shorter: a billion years, a million years, a thousand years, a few years, weeks, hours, seconds ... Clearly, this is bad news for life.

So, too much matter is bad. What if we go the other way, and reduce the amount of matter initially in the universe? For now, we'll forget about the existence of dark energy, and consider a universe with only matter.

With just slightly less matter, universes are born that expand so fast that they simply empty out. The density of dark matter and gas is so tenuous that atoms may be separated from each other by countless billions of light years. In these universes, there is no opportunity for structure to form, for gas to be funnelled into baby galaxies and stars, and, while they will live forever, these universes are eternally dark, lifeless places.

Thanks to the fine-tuning of the initial density of the universe, it doesn't take much to induce a suicidal expansion. If we look at the density of the Universe just one nanosecond after the Big Bang, it was immense, around 10^{24} kg per cubic metre. This is a big number, but if the Universe was only a single kg per cubic metre higher, the Universe would have collapsed by now. And with a single kg per cubic metre less the Universe would have expanded too rapidly to form stars and galaxies.

And as we push closer and closer to the Big Bang, the degree of fine-tuning gets larger and larger. The density of the Universe walked a cosmic knife-edge, and the smallest bump to the cosmic dials will erase the story of galaxy formation we told in the last section, and replace it with a much more boring tale. Dark matter will not collapse into haloes; ordinary matter will not collapse into disks; disks will not fragment into stars; the larger elements of the periodic table will not be made in stars and moulded into planets. The universe will be a nearly uniform hydrogen and helium soup, its atoms crushed momentarily under its own gravity or doomed to wander alone in infinite space.

Lost Horizons

About the same time that the flatness problem was identified, another problem emerged. Beginning in the 1970s, cosmologists started to grumble that the Universe was just too smooth!

Once we have solved Einstein's equations, we can trace not only the history of the matter in the Universe, but also the path of any particle of light moving through it. In particular, we can calculate which parts of the Universe could have communicated with each other by sending light rays. That's where the problems start.

Remember that the cosmic microwave background (CMB) is like an inside-out star, a shell of glowing matter around us. Put on a pair of microwave-detecting goggles™ and look in a particular direction into the night sky. You are seeing a particular patch of gas (call it *Patch A*) 378,000 years after the Big Bang. Now look just two degrees to your left on the sky, and call this *Patch B*. We can calculate that *Patch A* and *Patch B* are too far apart for any light to have passed between them in the 378,000 years since the Big Bang. They know nothing of each other. Just as ships on the ocean that are too far apart cannot see each other, even with perfect telescopes, because they are beyond each other's *horizon*, so we say that these patches are similarly beyond each other's horizon, a horizon set by the speed of light.

And yet, these patches of gas have the same temperature to one part in 100,000.

This is known as the *horizon problem*. We can see in the CMB millions of regions of the Universe that are outside each other's horizons. They cannot see each other. And yet, they are all almost identical.

Usually, a uniform temperature is a most unremarkable thing. The air in a room has roughly the same temperature throughout, and for good reason. Hot things make cold things hotter. And vice versa. If one part of this room were hotter than another, heat would flow until they reached the same temperature. All we need is contact.

In a universe that was always slowing down since the Big Bang, there was simply not enough time for these patches of the CMB to have any contact at all. We could simply assume that the Universe began with a uniform temperature, but this looks like another fine-tuning problem, another suspicious, unexplained arrangement.

The almost-uniform temperature of the CMB implies that the early Universe was almost smooth. The density of the Universe departed from the average density by at most one part in 100,000.

This number (1 in 100,000) is known as Q. We have no idea why it has this value, but this smoothness of our Universe is essential for life.

We saw in a previous section how these tiny lumps and bumps are the seeds of structure in the Universe, growing into galaxies. From these earliest stages, gravity was the driving force, with matter emptying out some regions that become immense cosmic voids, and flowing into regions that become the galaxies and clusters we see around us today.

Without these seeds, that story doesn't begin. If a universe began in perfect uniformity, then gravity would have nothing to grip. The universe would expand and cool, but the matter would remain forever smooth. In such a smooth universe, cosmic structure would simply not form; there would be no galaxies, no stars, no planets, and hence no life.

So a life-permitting universe cannot be too smooth after it was born; there must be some lumpiness to the distribution of matter and radiation.

In fact, Q is like Goldilocks sampling the bears' porridge; it must be just right.

Firstly, Q cannot be too small. Remember the important role that gas cooling plays in galaxy formation. Gravity is great but not enough – we need gas to radiate away its thermal energy if it is going to complete its final plunge into a galaxy.

By decreasing Q by a factor of 10 to one part in a million, ordinary matter remains diffuse, hot and unbothered. No dark matter halo is home to rapid, runaway gas cooling. The pressure of this hot matter holds gravity at bay, preventing collapse and fragmentation into stars. No stars means no planets and no life.

If we wind the value of Q upwards from one part in 100,000, what would be the effect? Clearly, with a larger Q, gravity has more to work with, and regions would condense and collapse in the early stages of the universe. Gas in these massive proto-galaxies would collapse and form stars, but more and more stars would be packed into smaller regions, with this close-packing resulting in many star–star close encounters, which would destroy the prospects of forming planets in stable orbits.

Taking Q to even larger values would result in gravity collapsing immense regions of the universe into extremely massive black holes, each weighing more than thousands of individual galaxies. In such a universe, all of the raw material for stars, planets and life would be locked away behind the one-way horizons of these black holes, and the universe would be dead and sterile.

How large does Q have to be before we create such catastrophic universes? Not large at all. If Q were one part in 10,000, then nearby stars would disrupt planets in their orbits. And if Q were one part in 100, then black holes would abound.

So, the value of Q, the initial lumpiness in the Universe, needs to be tuned to a high degree to ensure a reasonably sedate beginning to the formation of structure, but not so sedate that it does not occur at all.

Cosmic Inflation

We've left a few loopholes, and a few *ifs* and *supposes*.

The flatness problem and the horizon problem follow from assuming, amongst other things, that matter and radiation dominated the energy of the Universe in its earliest stages. The Universe's deceleration drives it away from flatness, and isolates the patches of the microwave sky.

This suggests a solution: a period of accelerating expansion in the very early Universe. This idea, which was developed by Alan Guth, Andrei Linde, Alexei Starobinsky, Paul Steinhardt and Andreas Albrecht in the late 1970s and early 1980s, is known as *cosmic inflation*.

To solve these problems, inflation must have been extreme. Beginning roughly 10^{-35} seconds after the initial birth of the Universe, inflation lasts until 10^{-34} seconds. In that time, the Universe doubles in scale at least 80 times (about a trillion trillion times!). Think of inflating a grain of sand to the size of our galaxy, more than 100,000 light years across.

Though it lasts but a moment, the acceleration of inflation smooths out the curvature of space, leaving the Universe almost perfectly flat. It also ensures that our microwave sky shows only a small patch of the pre-inflation Universe. In particular, we see a patch small enough to have reached the same temperature.

Inflation gives a bonus prediction. Because of the quantum jittering of both matter and space itself during inflation, the Universe is not perfectly uniform. It is seeded with structure, the small lumps and bumps that (as we have seen) will go on to form galaxies. The novelty of inflation is that, with a few mathematical twists and turns, it can predict the statistical properties of these fluctuations (not just their size, Q), and can get them right.

Are we done? Fine-tuning solved? After all, we have a testable, successful scientific theory that explains the fine-tuning of cosmological parameters. Don't we?

Not so fast. The problem with the theory of cosmic inflation is ... it's not really a theory. There's no physics.

Inflation is defined as a period of accelerating expansion. As we have seen, the expansion of a universe (accelerating or otherwise) is caused by the energy contents of the universe. So, what causes inflation? The idea itself doesn't say.

Inflation is an effect, not a cause. It is an effect that produces other, desirable effects. But there are hundreds of ideas of what could cause inflation, with little to separate the candidates.

As an analogy, inflation is like the white ball in pool. If set on the right course it can pot an object ball neatly in the corner pocket. But that's not a full explanation of the pot: we're missing the cue!

So, what kind of energy could cause inflation? It's probably some form of field, and since fields are named after their associated particles, it's usually called the *inflaton field*; this impressive sounding object is just another we're-in-the-dark name.

Any physical theory of inflation needs at least six things.

A. We need a form of energy with negative pressure. As we saw in our discussion of dark energy, negative pressure will cause the expansion of the universe to accelerate. The cause of inflation cannot be dark energy, as it simply lacks the ability to power such rapid expansion. So, we need yet another, unknown form of energy.

B. Inflation must start. The inflaton field must, at some point, control the universal expansion. We won't get accelerating expansion if the energy of the inflaton is swamped by the energy of matter and radiation.

C. After starting, inflation must continue. To solve the flatness and horizon problems, we need at least 80 doublings of the scale size of the universe. Inflation needs to slam the accelerator pedal!

D. After continuing, inflation must end. This is an important one! During inflation, matter is diluted by a factor of at least a trillion trillion trillion trillion trillion trillion, and at the end of inflation there is nothing to form cosmic structure of *any* kind, let alone life.

E. After inflation, for there to be any hope for cosmic structure, the universe must be refilled with matter and radiation. As suddenly as it appeared, the inflaton now has to vanish, but as it does so, its energy must be converted into ordinary matter and radiation.

F. As matter and radiation flood back into the universe, they mustn't be too smooth or too lumpy. Recall from our discussion above that life needs a very specific amount of lumpiness: Q between one part in 1,000,000 and one part in 10,000.

Without knowing what the inflaton is, it's hard to judge whether these requirements are generic or fine-tuned. However, we have a few clues.

The easiest one is probably C: once inflation starts, it's not hard to maintain. We have some reasonable clues about how to accomplish D and E. While early ideas about inflation struggled with the 'graceful exit problem', newer theories can get the job done. Except ... the inflaton field is yet another quantum field that is expected to contribute to the vacuum energy of the universe. If, once inflation is over, its vacuum energy is too large (positive or negative) then it will either continue accelerating the expansion or rapidly end the universe in a crunch.

A is perhaps worse than dark energy – it can't be *just* vacuum energy. Vacuum energy doesn't stop. So we're even more in the dark about the inflaton. We have a similarly large number of auditioning candidates, and few clues as to who should be cast.

B is a major problem. It is still a matter of some debate among cosmologists whether inflation can be expected to take centre stage in any old universe, or whether inflation itself requires special, fine-tuned initial conditions. For example, many of our ideas require the inflaton to control a large, smooth region before inflation even starts. But a large smooth region is exactly what we wanted inflation to explain. Similarly, a field that can produce inflation looks like a low-entropy state. In that case, inflation only succeeds in replacing one fine-tuning with another.

Finally, F is the 'famous fine-tuning problem of inflation', according to Professor Neil Turok of the Perimeter Institute for Theoretical Physics (2002). Inflation can produce practically any value of Q, from zero to very large values. If Q is greater than one, the universe comes pre-loaded with black holes; this really is not a good idea. The properties of the inflaton must be fine-tuned to produce the right value of Q, so again we replace one fine-tuning with another.

So inflation makes the cosmological constant problem worse, and seems to need fine-tuning to solve the flatness and lumpiness (Q) problems. While inflation is an impressively predictive idea, and may indeed explain much about our Universe, we aren't done yet.

The Impact of the Humble Neutrino

There is another cosmic dial that life must keep an eye on. Looking back at the zoo of particles that make up the Standard Model, there are three particles that we haven't much discussed: the neutrinos.

Neutrinos (with symbols v_e, v_μ and v_τ) are created in nuclear reactions, but they are notoriously difficult to detect. Because they do not carry any charge, they do not interact with other charged particles via the electromagnetic force. So, when they are produced in nuclear reactors, be it in a nuclear power station on Earth or in the reactions at the heart of the Sun, they rapidly stream through matter and off into space.

Talking about the mass of the neutrino was, for the longest time, quite a simple affair. Since the prediction of their existence in the 1930s, experiments showed that neutrinos appeared to possess no mass whatsoever. Neutrinos were massless, like photons of light, except that we cannot see them. Boring light. How wrong we were!

Our story begins at the heart of the Sun. The nuclear reactions that power the Sun, as well as being a source of light, are a prodigious producer of neutrinos. In fact, almost 100,000,000,000 solar neutrinos stream through each square centimetre of your body every second.

We can feel how much energy the Sun emits by simply sunbathing on a summer day, and with some simple equipment we can find that more than a thousand joules of energy hit every square metre of the Earth at midday. From this, and a bit of physics, we can deduce how rapidly the nuclear furnaces at its core must be burning, and from this we can deduce the expected flux of neutrinos at the Earth's surface. While neutrinos are difficult to detect, they are not impossible, and over the last 30 years we have been able to measure the flow from the Sun. And there was a problem. A big problem!

Scientists counted the neutrinos arriving at their detectors and calculated that there was only one-third the expected number. Something was badly wrong. Theorists scratched around within the laws of nuclear physics, and there appeared to be no problem there.

In the Standard Model of particle physics (Figure 9), there are three types of neutrinos, one for each of the massive leptons. One of these is associated with the electron, and one each for its heavier siblings, the muon and the tauon. The nuclear reactions in the Sun predominately produce just one of these, the electron-neutrino, and the detectors on Earth were constructed to measure the flow of this species of neutrino passing through it.

So, with the neutrinos streaming from the heart of the Sun, what if, instead of them being all electron-neutrinos, some were those of the other flavours? What could possibly do this?

The answer is not in the nuclear physics that powers the Sun, as we know that this produces only electron-neutrinos, so *something* must happen on the journey from the Sun to the Earth. And to understand this, we are going to have to call on some of that weirdness that occurs in the world of quantum mechanics (fear not; the quantum world is only really weird if you insist on comparing it to everyday experiences).

What happens to the electron-neutrinos is called *quantum mixing*. While travelling, the electron-neutrino can change its spots and become a muon-neutrino or a tauon-neutrino. This spot changing is cyclical, and, like fickle changes of fashion, the neutrino could find itself as an electron-neutrino again.

This mixing occurs all along the journey's path, and while the Sun creates 100 per cent electron-neutrinos, by the time they reach the Earth, only 33 per cent of them still are that type, with the other 67 per cent invisible to the detectors that were built to register the passage of electron-neutrinos. Problem solved!

The mixing of neutrinos reveals a very important fact about them: neutrinos have mass. A very small mass, admittedly, but mass nonetheless. Perhaps the shadow Universe is unimportant.

In fact, the small mass of neutrinos could make a big difference to the Universe. Why? Because they are so numerous! We've already seen that neutrinos are formed in all of the nuclear reactions in all of the stars in the Universe, but they are also emitted during some radioactive decays. More importantly, neutrinos were produced in abundance in the initial moments of the Big Bang.

In fact, if we add up all of the various sources of neutrinos in the Universe, there are, on average, about 340 million per cubic metre. This completely swamps the average density of atoms in the Universe, which is only about two hydrogen atoms per cubic metre, and is roughly equivalent to the density of leftover radiation in the cosmic microwave background.

It is rather sobering to think of this shadowy Universe, coexisting spatially with our own, but hardly felt. But there is one place that neutrinos make their presence known, and that is in their gravitational pull.

When the Universe was young, before the first stars were born, it was a sea of radiation, matter and neutrinos. Gravity was seeding galaxies, with collapsing dark matter pulling in the gas that eventually formed into stars and planets. In our Universe, nearly massless neutrinos play a very small part. Their energy is overwhelmingly in the form of kinetic energy, and so they move at essentially the speed of light. They stream quickly away from any lump of matter, refusing to be bound and add to its attractive pull.

But, as with the other fundamental particles, we have no idea why neutrinos have the masses they do, or why their masses are so small when compared to their fundamental cousins. And with the other particles, we can ask: how different would the Universe be if the neutrinos had a different mass?

In fact, this question has been investigated in great detail by cosmologists. By understanding how the mass of the neutrino affects how structure in the Universe grows, we can use our observations of the structure of the Universe to measure the masses of the neutrinos! (This is a great illustration of the continuity between fine-tuning in particular and theoretical physics in general.) Impressively, these measurements of the neutrino mass are some of the best available: the sum of the masses of the three types of neutrino is less than a millionth of the mass of the electron.

Max Tegmark, Alexander Vilenkin and Levon Pogosian (2005) put this marvellous physics to work, calculating the influence of more massive neutrinos on the formation of galaxies. If the neutrinos were substantially more massive, then this would make them more sluggish, and they would struggle to stream away from newly formed galaxies. As they hang around, their mass adds to the total mass of the baby galaxy, and so will help it grow.

But what about a smaller increase in the mass of the neutrino? These neutrinos would still be nimble enough to stream out of the birth sites of galaxies and fill the universe. You might think that that's the end of the story, with the cosmos possessing a sea of slightly more massive neutrinos. But this extra mass spread throughout space has a cumulative effect. Galaxies grow through the accretion of mass over time, and how fast mass can be accreted depends upon the contrast of the local density at the site of the baby galaxy to the background density of the surrounding universe. The extra mass spread out in the form of massive neutrinos significantly reduces the contrast, reducing the rate at which galaxies grow. The net result is that slightly boosting the neutrino's mass inhibits the collapse of matter into galaxies, and without this collapsing material we do not have the power sources to generate life, the stars, and planets on which life can emerge and evolve.

But how big a change can we make? Well, not that much. As we've seen, neutrinos are about one millionth of the mass of the electron, the next most massive particle. Tegmark and collaborators showed that increasing the neutrino mass by even a factor of a couple will have a devastating effect on galaxy formation, effectively suppressing it completely and leaving the universe nothing but a smooth soup of shapeless matter.

So, for complex life to arise on the surfaces of planets orbiting stars in galaxies, a universe needs an almost massless neutrino. Why nature obliged is still a mystery.

COSMOLOGY AND MYSTERY

For the moment, the fine-tuning of cosmology remains. But cosmology is not short on mysteries. The laws of physics have allowed us to investigate, extrapolate and speculate about the history of the Universe from about 10^{-35} seconds after its apparent birth to today, roughly 13.8 billion years later. This is pretty impressive!

In our quest for a deeper explanation of fine-tuning, we need to push back even further, to the very beginning. But here, at about 10^{-43} seconds, we eventually run out of speculation. Our best theory of gravity – Einstein's General Relativity – breaks down. Finally, at $t = 0$, the creation of the Universe, we hit a mathematical absurdity called a *singularity*. If we ask Einstein's theory, what was the density and temperature at the birth of the Universe, we get the answer 'infinity'.

In much of popular science, singularities are often presented as being a mysterious, exotic kind of object, but in physics, the appearance of infinities is a red flag, warning us that we have made an unrealistic assumption. We have pushed our equation beyond its limits.

It is important to remember that a singularity is not an object, and especially not a magical one; it is simply a mistake resulting from pushing our theories too far. Just like mistakes in everyday life, such as spilling the milk or crashing a car, singularities should not be talked about in hushed, reverent tones, but rather examined to determine what went wrong.

And yet, the singularity at the beginning of the Universe is particularly stubborn. Our simplest, and yet most successful, mathematical theories of the Universe all begin with a singularity. Adding more realistic lumps and bumps has been tried, but to no avail. In fact, in the late 1960s, Stephen Hawking and Roger Penrose showed that in all sorts of reasonable universes, a singularity at the beginning is unavoidable.

Cosmologists have continued to ponder the beginning of the Universe, uncovering a variety of competing and confusing hints, including bouncing universes and singularities erased by quantum fuzziness.[12] Here, then, is cosmology's frontier, its wild west. Progress will undoubtedly be slow and painful. But our quest for answers has left us staring at what looks like a beginning, but is just out of reach. We can only push on.

---

[12] Quantum mechanics: is there anything it can't do?!

SOME FINAL THOUGHTS ON INFLATION

How Unusual Is Flatness?

We cannot leave inflation without noting a peculiarity. It has to do with the flatness problem: why is our Universe, after a few billion years of evolving away from the critical, flatness-inducing density, still so close to it? The early Universe has a wide range of densities to choose from; why pick such a special one?

One problem is that *all* universes dominated by matter and radiation in their earliest stages start out very nearly flat.[13] The only difference is how long it takes for the not-exactly-flat ones to evolve away from flatness. Ours shows no such signs after 13.8 billion years. Should we think of that as a long time or a short time? While it is obviously long on human timescales, classical cosmology gives us no standard time to compare. And so our question's answer is: 'long or short compared to what?'

From quantum gravity, however, we do get a standard time. We might expect the classical initial conditions of cosmology to be laid down at the Planck time, at 10^{-43} seconds, when gravity and quantum mechanics part ways. So, it is not true that *all* universes start out flat, since (rewinding the clock) some of them don't reach flatness by the Planck time. In that case, the flatness problem is equivalent to asking: why is the lifetime of our Universe so much longer than the Planck time?

But a second peculiarity is lurking. Whatever time we pick, we are faced with a range of densities. Are all these other densities really as available as we think? Being good scientists, we'd like to say how probable a given universe is, given our theories. Given such a narrow range, we might guess that the probability of a very-nearly-flat universe (without inflation) is very small.

However, there is a more rigorous way to do the calculation. It's rather technical, not easily computed and gives a surprising result. The measure of the exactly flat universes is 100 per cent.

---

[13] This point is well made by Helbig (2012).

A probability is a type of measure, but a measure is not necessarily a probability. You can measure the area of the bullseye relative to the whole dartboard, but that only becomes a probability if you start throwing darts. But this result, first derived by Stephen Hawking and Don Page, suggests that flat universes are utterly unsurprising.[14] In fact, non-flat universes are the unusual ones.

This is a strange result, and one that is difficult to interpret in isolation from some particular scenario for the beginning of the Universe. What does this measure mean? If it's a probability, what is it the probability *of*? Is there a statistical sample of universes? Is it simply a summary of what we should expect to be true?

Whatever the case, we have reason – independently of inflation – to view the flatness problem with some suspicion. The total energy density of the universe needed by life might not be as fine-tuned as first appearances suggest. This does not affect the other problems to be solved by inflation, such as the horizon problem.

Inflationary Target

Does inflation show the way forward? Could other fine-tuning cases be solved in the same way?

Probably not, because inflation has a target.

Think about packing energy into a young universe. Pick a particular time, and lay out all the possible energy densities – the energy you're going to squeeze into each cubic metre – on a number line. On the left, at zero, is a completely empty universe. Out to the right is the Planck density – to know what happens further to the right, we would need a theory of quantum gravity. So 'here be dragons'.

Now, do Einstein's equations pick out any point on this line as unique? Yes – given a certain rate of expansion, there is one particular density that separates negatively curved universes on the left, positively curved universes on the right, and flat universes right on the fence. This is the critical density.

[14] Hawking and Page (1988). See also Evrard and Coles (1995), Gibbons and Turok (2008), and Carroll and Tam (2010).

Now, what density does life need? The range of values that permits the formation of cosmic structure – galaxies, stars, planets and people – sits right on the fence.

This is how inflation (and the measure from the previous section) *could* solve the fine-tuning problem. They have a target. Aim for flatness, and you get life thrown in for free.

But this is precisely what is missing for almost all other fine-tuning cases. Draw the number line for the quark masses, for the electron mass, for Q, for the strengths of the forces – in none of them will you find a target, a special value waiting inside the life-permitting region. There is nothing at which to aim; physics is blind to what life needs. And yet, here we are.

6 All Bets Are Off!

For all this talk of changing the laws of nature, until now we have stayed safely within the confines of familiar physical theories. We've messed with the ingredients in the laws: the stuff, the strengths of the forces, and the initial state of the Universe. But we've not touched the laws themselves. However, just like the constants of nature, we don't know why the laws of nature are what they are. So let's change *everything* ... all bets are off!

You may think that we've already discussed some bizarre scenarios thus far, but this is where things get very weird. We will start in a gentle fashion, looking at the division in our Universe between familiar classical physics and quantum mechanical weirdness. We will then delve deeper and look at the role of symmetry in defining the properties of the cosmos. This leads us to alter the very fabric of the Universe, space and time. Does life have a chance in any of these unfamiliar arenas?

Before we dive in, a note of sanity. We have so far emphasized that, ultimately, we don't know why things are the way they are. At the cutting edge of science, countless crazy theories float freely, unhindered by experimental evidence. To the layperson, this chaos can seem aimless, and it can be difficult to discern what is profound and what is pure fiction.

But the scientist loves a bit of chaos. If the chaos is theoretical – when we have too many ideas and not enough evidence – then a decisive experiment can cut a swathe through our theories, excising the debris, and greatly advancing our understanding of nature's ways. If, on the other hand, the chaos is experimental – we have too many unexplained observations – then theoretical physicists must think deeply and creatively, and be prepared to advance bold new ideas.

The history of science is the story of clarity gradually emerging from confusion; from the Copernican revolution – the Earth goes round the Sun! – to quantum mechanics in the early 1900s, to the discovery of DNA in the 1950s, chaos leads to insight. The confusion swirling around cosmology, in which we don't know the identity of 95 per cent of the Universe, means that the time is ripe for a dramatic shift. We will, therefore, not apologize for dwelling on what we don't know. Rather we celebrate it! You should too![1]

ON THE SHORES OF THE QUANTUM OCEAN

From the early days of modern science, through the eyes of Galileo and Newton, science was focused on everyday experience. While not mathematically simple, as a glance at a classical mechanics textbook will demonstrate, physics maintained a very direct connection with the world we experience. Newton explained how apples fall, how the planets move, and why rainbows are seen on a showery day. Two hundred years later, the equations of James Clerk Maxwell united electricity and magnetism, and revealed the nature of light. Fired by the industrial revolution, the science of thermodynamics had unravelled the mysteries of heat, temperature and pressure.

'Almost everything is already discovered', a young Max Planck was told in 1874. Planck, who would become one of the greatest scientists of the twentieth century, had travelled to Munich to embark on a career in physics, only to be told by Professor Philipp von Jolly to study something else, as 'theoretical physics was approaching a degree of completion which geometry had possessed for hundreds of years'.[2]

Other physicists knew better. Cracks were appearing.

---

[1] While you should celebrate the chaos at the cutting edge of science, you should always take any media stories associated with it with a rather large pinch of salt. We actively support the media including as much science as possible, and dream of the day when the news will carry a science report just before the sport and weather. However, *breakthrough fatigue* is an occupational hazard of any reader of science journalism: every day, it seems, another press release proclaims an idea that will change the way we see the Universe, but they never quite live up to their promise.

[2] Quoted in Weinert (2004, p. 193).

Atoms were on physicists' minds, and in particular how their innards were arranged. The emerging picture resembled a miniature Solar System (Figure 8), with the nucleus at the centre and the electrons orbiting like planets. But there was a problem, a big problem! Electrons are held in their orbits by the pull of the electric force towards the atomic nucleus, in the same way that the Sun's gravity pulls on a planet. But, unlike the gravitational force, orbiting electrons should lose their energy by emitting radiation and, in the blink of an eye, crash into the nucleus. Something is wrong with the Solar System atom picture; it can't explain why atoms are stable!

But unstable atoms were not the only difficulty for classical physics. Nothing in the nineteenth-century physics textbook could explain the light given off by iron heated in a forge, or the electrons spat out when light was shone onto a piece of metal such as potassium.

After several decades of confusing experiments, false starts, dead ends, crazy ideas, cruel and unusual mathematics, and a few moments of true genius, it became clear that the rules of the microscopic world are not the same as those of everyday objects. In the microscopic world, quantum mechanics rules.

To explain the properties of the quantum world properly would take another book (and then some), but we can summarize the important points. In classical physics, particles are represented at any particular time by their position and velocity; that is, where they are and where they're going. In quantum physics, it's a bit more complicated. Particles such as electrons have wave-like properties, so we describe them with a *wave function* whose peaks and troughs tell where the electron *probably* is, and where it is *probably* going.

When an electron orbits the nucleus of an atom, it cannot simply wander wherever it pleases. In quantum mechanics, the electron is a wave, and must fit snugly into its orbit. Figure 32 shows electron waves around a nucleus; on the right is an orbit that is not allowed – the ends of the wave don't join up. The other waves fit neatly together, and show that there is more than one way for the electron to orbit. Like the energy of a photon, the energy of an electron in its orbit

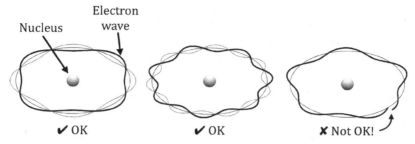

FIGURE 32 An electron around a nucleus must fit its wave into the orbit. Thus, certain waves (and their associated energies) are allowed, such as the left and middle figures, and others are forbidden, such as the orbit on the right.

depends on its wavelength: the more tightly packed the wave, the higher the energy. Because not all waves are allowed, the electron's energy is *quantized*; that is, it can take on *only* discrete values.

Why suppose that electrons have these strange, wave-like properties? Because we can understand why atoms are stable! There is an *innermost* orbit: the electron can only get so close to the nucleus; that is, close enough to fit in one of its waves, and no closer.[3] From this *ground state*, the electron cannot fall closer to the nucleus because there is no allowed orbit into which it can move.

Thus, while the quantum world has a common reputation for being fickle and unpredictable, it is actually the basis for the stability of all the matter you see around you.

While strange, physicists have come to accept that this is the way that the world of the very small works. The astonishingly accurate predictions of quantum mechanics have underpinned some of the

---

[3] Here is a slightly more sophisticated version of this argument. The electron's wave nature places a limit on how confined its position and momentum can be. This is Heisenberg's famous *uncertainty principle*. An electron spiralling into the nucleus will have a smaller and smaller uncertainty of position, and so a larger and larger uncertainty in momentum. At some point, the electron is too close, and so moving too fast, to be confined by the electromagnetic force. The competition between the confining attraction of the electromagnetic force and the quantum resistance to such confinement sets the size of an atom.

great technological inventions of the modern world, from the computer chip to the laser to magnetic resonance imaging (MRI).[4]

But, where is this quantum waviness in everyday life? Why does it affect atoms but not tennis balls? The physicist has two sets of equations, one set that we would use for an electron flying across a room, and another for a tennis ball flying across the very same room. How do they fit together? And when, where and how do quantum things manage to look like classical things? Where is the *quantum–classical transition*?

The answer seems to be tied up with the value of another constant of nature, Planck's constant. This number, via its symbol h, appears in all of the key quantum mechanical equations. It is there when you want to calculate how particles deflect as they travel through narrow slits, the size of electron orbits about the nucleus, the structure of atomic nuclei, and the emission of light from the hydrogen atom.

In our Universe, Planck's constant has the value $6.626\ 069\ 57 \times 10^{-34}$ kg m^3 s^{-2}. This number, measured in human-sized units of kilograms, metres and seconds, is very small. The size of an atom is proportional to Planck's constant, implying it must be small; there are about 10^{24} molecules in a glass of water. It also tells us that the energy of an individual photon is miniscule; a typical light bulb emits around 10^{20} photons per second. Because human beings are too big to see this fine-scale granularity, water and light are effectively continuous.

So, can we simply say that big things are classical and small things are quantum mechanical? Can we avoid quantum weirdness in everyday life by banishing it inside atoms? This won't quite do, as a famous – and infamous – thought experiment by Erwin Schrödinger shows.

Schrödinger, in 1935, considered the following 'diabolical device'. A cat is imprisoned in a box with a vial of poison and

---

[4] Unfortunately, the word *quantum* has been hijacked to mean *magic* and has been used to explain all kinds of 'phenomena' from telepathy to ghosts. Putting *quantum* on a health product will not turn snake oil into a panacea.

a radioactive atom. In the next hour, there is a 50 per cent chance that the atom will undergo radioactive decay, triggering a detector that smashes the vial and kills the cat. The box is sealed, and we have no knowledge of whether the vial has been smashed or not, and so do not know if the cat is alive or dead. We wait for one hour.

Now, before we open the box, what should we say about the cat? Schrödinger's cat is often presented to the public as a philosophical conundrum. 'The cat is both alive and dead, according to modern physics', the increasingly confused layperson is told. Now we'll scratch our collective chins, look thoughtful, and try to convince ourselves that this is somehow profound.

However, that is *not* the point of the thought experiment. In the mathematical language of quantum mechanics, the answer to the question 'has the atom decayed?' is *'atom intact + atom decayed'*. What on Earth does that mean?

Such a state is called a *superposition*. What we should *not* say is that 'the atom *both* has decayed and is intact'. In quantum mechanical language, there is a perfectly precise way to say 'the atom has decayed', and a way to say 'the atom is intact', and the superposition is *neither* of these. Rather, if we think that the wave function represents the system itself, we should say that it is the case that the atom has neither decayed nor is intact.[5] On the other hand, if we think that the wave function represents our knowledge of the system, then we say that we don't know whether the atom has decayed or is intact.

But never mind the atom's microscopic indecision, the usual story says. It will be ended when the detector registers the decay.

But wait ... the detector is a quantum system, too. It is made of atoms. When we take this into account, quantum mechanics tells us that the entire 'atom and detector' system is described mathematically

---

[5] We need not deny logic's 'law of excluded middle': either A or not-A is true. If A is 'the atom has decayed', then not-A is 'it is not the case that the atom has decayed'. A superposition 'decayed + intact' is the latter case, not-A. In quantum mechanics, 'it is not the case that the atom has decayed' does not imply that 'it is the case that the atom is intact'. The superposition 'decayed + intact' is something else entirely.

as 'decayed & detector triggered + intact & detector untriggered'. The wave function is still undecided.

But, as the atom and detector interact with their surroundings, more things become *entangled* with the atom's indecision, including the cat. The system is soon described as 'decayed & detector triggered & cat dead + intact & detector untriggered & cat alive'. Remember, this means that the wave functions of all the atoms in the cat and detector are entangled with the original atom's indecision. What a mess!

Note well what has happened, for this is Schrödinger's point, a point often lost in the discussion of his unusual cat eradication device. The atom's indecision is now cat sized! The quantum 'neither here nor there' weirdness of superposition has been amplified, so that we are forced to say that the cat is a superposition of 'alive + dead'. But, says Schrödinger, this is 'quite ridiculous', and never observed! The point of this thought experiment is not that we must ponder how a cat can be both alive and dead, but instead to note that we never observe such a thing. When we open the box, we do not see a mixed, alive-and-dead cat. We do not ourselves become a blurry blend of 'I see a living cat + I see a dead cat'. We see one or the other. In quantum mechanics, we are presented with superposed, entangled alternatives. But the classical world – the world we observe – presents us with a single reality, with alternatives that are merely possible, not superposed.

So, it's not as simple as 'small is quantum, big is classical'. We're still no closer to answering our question: when, where and how do quantum things manage to look like classical things?

Part of the answer seems to be a phenomenon called *decoherence*.[6] Recall that every particle that interacts with the 'atom + detector + cat' system becomes entangled in the atom's indecision. Because the atom can't decide what to do, neither can anything that interacts with the atom. (This is metaphorical language, of course. Atoms don't 'decide'!)

---

[6] Note that we will be playing a little fast and loose with our analysis of the transition between classical and quantum mechanics.

However, we cannot possibly keep track of every particle in our experiment. A photon can bounce off the detector, out of the window and off into space. In practice, we keep track of a *subsystem*.

Here's where *decoherence* comes in. All the quantum weirdness and waviness is only seen when we can keep track of the whole entangled system. This is much easier to do if the system is small. But, if the system has interacted with its environment – its big, messy, untracked environment – then the quantum weirdness goes away. It leaks into the surroundings. What remains, astonishingly, acts like a classical system!

More precisely, it is because our measuring devices (such as the detector, the cat and your eye when you open the box) don't keep track of all the quantum entanglements that we fail to see quantum weirdness in large systems. Quantum subsystems, by interacting continuously and irreversibly with their environment, are forced into those special states that act classically. Decoherence destroys superpositions.[7]

Well, almost. We still have to explain why only *one* of the possible classical states is observed. Decoherence is part of the answer, but quantum mechanics is still missing something. We cannot discuss all the options here; we can report that they are all weird.

Our increased understanding of decoherence has meant that scientists have learned how to amplify quantum weirdness without destroying it. We have been able to create and observe large molecules, with 800 individual atoms, in superposed states. These molecules are approaching the size of the smallest life forms. Soon, an amoeba, or worm, or maybe even a mouse, may *experience* quantum weirdness.

But we should return to fine-tuning. How different would the Universe be if we messed about with Planck's constant, and changed the scale where quantum mechanics becomes important? Imagine if we could set Planck's constant to zero. We would effectively banish the influence of quantum mechanics, and the world would obey the purely classical laws of Newton and Maxwell. As we have already

---

[7] For more technical details about decoherence, see Zurek (2002) and Dass (2005).

noted, in such a classical universe, atoms are unstable as electrons lose energy and spiral into the nucleus. In a universe with no quantum mechanics, a universe with $h = 0$, there are no atoms or chemistry, and hence no molecules or people.

Suppose, instead, we changed Planck's constant to increase the scale at which quantum weirdness showed itself, until superposition affected large, *macroscopic* objects.[8] That is, what if decoherence was so slow that most objects spent most of their time in quantum superpositions?

We are not the first to consider this question. Mr Tompkins, the protagonist of a series of popular science books by the Russian physicist George Gamow, encounters balls on a pub billiard table behaving strangely, spreading out as they roll until 'it looked as if not one ball was rolling across the table but a great number of balls, all partially penetrating into each other. Mr Tompkins had often observed analogous phenomena before, but today he had not taken a single drop of whisky.'[9] The billiard balls behave like a 'peculiar wave' after a collision and leak out of the wooden triangle. Later, having ventured into a quantum jungle, Tompkins is attacked on all sides at once by a quantum tiger! How do we understand this strange quantum world?

In the classical world, physics is *deterministic*, meaning that if we know exactly how the world is *now* – we know where all the atoms are and where they are going – we can use the laws of physics to work out what the world *will be*.

But the quantum world is quite different. Knowing the quantum state of an electron doesn't necessarily tell us where it is (location) or where it's going (momentum), and it certainly can't tell us *both* its precise location *and* momentum. We still have equations that allow

---

[8] Recall that h has units, and our units (seconds, metres, kilograms) are bound up with quantum physics. For example, the second is defined using oscillation of a caesium atom, which in turn depends on its quantum properties, and hence depends on h. So, to properly specify a different universe with a different value of Planck's constant, we should make explicit what quantities are different and what stay the same. Here, we'll simply consider the general consequences of life-sized quantum effects.

[9] Gamow (1965, p. 65).

us to evolve these properties from one point in time to another, but this quantum fuzziness remains.

Consider a life form that had to deal with this quantum uncertainty directly. It would be like Schrödinger's 'quite ridiculous' cat, never knowing where things are or where they are going. Even more confusingly, its own body could be similarly fuzzy. How could you operate in a universe where you could be walking, sleeping, dancing and/or diving all at the same instant? How could your brain store and process information if it spends most of its time unable to make up its mind? (So to speak![10])

A particularly thorny problem in a quantum universe is *non-locality*. Consider a ball that has been kicked. When predicting its flight path, whether this is done mathematically by a physicist or intuitively by a footballer, we need only consider the effect of objects near the ball. So we consider the wind speed in the stadium, but not the wind speed over the Atlantic Ocean. But in a quantum world, objects aren't so neatly separated. They can be *entangled*, and so in principle the ball's flight could be affected by anything else in the universe. To us, with our finite knowledge of the world around us, the ball would be radically unpredictable.

So, in our Universe, we have quantum mechanics just where we need it, influencing only the very small. It stabilizes atoms, but cats couldn't care less about the quantum pieces from which they are made.[11] We can rely on our predictable, classical world. In particular, life can store and process information, which is one of its defining qualities.

UNIVERSAL SYMMETRY

There are other, more radical changes we can make to the laws of physics. We need to talk about symmetry.

---

[10] For the reader who has heard of quantum computers . . . those won't help. A quantum calculation ends with a 'measurement'. That is, quantum computers need decoherence (at least) to finish the calculation. In a universe in which decoherence is very slow, completed quantum calculations would be the exception rather than the rule.

[11] Cats don't seem to care much for Newtonian mechanics either.

Like *quantum, symmetry* is a term that sounds like it belongs in a discussion of home decorating, feng shui and crystals. In fact, symmetries are extremely important to modern physics. They are both practical and profound, bringing simplicity and beauty to our equations. To many physicists, the search for ultimate laws of physics simply is a search for the deepest symmetry.[12] And, importantly, from symmetry we get *conservation laws*.

Here's how it works: wherever a symmetry can be found in the structure of the Universe, you will find an associated *conserved* quantity that doesn't change with time. To understand this, let's look at some examples.

From high-school physics you may remember a host of conserved quantities, things like energy, momentum and electric charge. The total amount of these things remains the same over all time; energy, for example, cannot be created or destroyed, but only converted from one form to another. Other quantities are not conserved. In classical physics, while the total amount of energy is conserved, kinetic energy alone isn't; cars turn petrol into motion, and then motion into heat as the brakes are applied. Magnetic fields, temperatures, gravitational acceleration and many other physical quantities are not conserved either. Conserved quantities are much beloved by physicists because they need only be calculated *once*. Whatever the messy predicament in which your system finds itself, the conserved quantities will not have changed.

We have the brilliant mathematician Emmy Noether to thank for linking symmetries with conservation laws. Noether is an unsung hero of modern science, an intelligent woman at a time when academia was completely dominated by men. 'In the judgment of the most competent living mathematicians, Fräulein Noether was the most significant creative mathematical genius thus far produced since the higher education of women began,' noted Einstein in her *New York Times* obituary.

---

[12] We heartily recommend Steven Weinberg's *Dreams of a Final Theory*.

Here is a simple example of Noether's insight. Imagine you do a physics experiment in your lab, and you get a particular result. You don't expect to get different results if you move all your lab equipment one metre to the left, because you assume the laws of physics over there are the same as here. Sounds obvious, even trite, but using a particularly powerful approach to classical physics known as *Lagrangian mechanics*, this *translational* symmetry implies the conservation of momentum.

Similarly, any physics experiment done today gives the same result as an identical experiment conducted yesterday, or tomorrow. This *time* symmetry, thanks to Noether, implies the conservation of energy.

But there is more to physical things than where and when they are, and thus there are more symmetries to be found, though they will not be as easily visualized. For example, a symmetry hidden within the quantum mechanical wave function results in the conservation of electric charge.

The link between symmetries in nature and the existence of conserved quantities often comes as a surprise to students of physics. But once you realize how it works, hunting symmetries and their conserved quantities is a powerful way of proposing new laws of nature. Every conserved quantity not only makes solving problems a bit easier, it can help us find the right problem in the first place.

But, we have to be careful. Just as physical theories have limits, so too do symmetries, and conservation laws that appear quite obvious in some circumstances may not hold in general.

Energy, the most famous of the conserved quantities of physics, is an interesting case. Remember that it is related to a symmetry of spacetime, specifically that our laws do not depend on time. But, as you may recall from the previous chapter, Einstein's General Theory of Relativity meddles with spacetime itself, and so need not display any spacetime symmetries at all!

We see that energy is conserved in experiments we have performed on Earth, so time translation symmetry appears to hold in our cosmic neighbourhood. But we also know that the Universe as a whole is expanding; the distance between any pair of galaxies will be larger tomorrow. If the Universe is finite in size, then tomorrow it will be bigger in the sense of literally having more space.

The fact that the Universe is expanding means that the space-time *itself* does not possess time symmetry. So ... wait for it ... the Universe as a whole does not conserve energy!

Gasp! What scandal! 'But what about all that stuff I learnt at school about not being able to create or destroy energy?' you might be wondering. Well, school got it wrong.[13]

We can see this non-conservation in action. Light, for example, is redshifted to longer wavelengths as it travels through an expanding universe. Each photon loses energy, but where did it go?[14] Answer: it doesn't go anywhere. The Universe does not obey time translation symmetry, and so energy is not conserved. Dark energy also refuses to conserve energy; if the Universe doubles in volume, then it contains double the amount of dark energy.

This doesn't mean that chaos reigns. The energy of the Universe changes predictably, so we can follow how it flows and transforms. In fact, understanding how energy in a rapidly expanding Universe dictates the forging of the elements in the first few minutes has led to some of the most impressive predictions of modern cosmology.

So we live in a Universe that doesn't conserve energy. If this makes you a little uncomfortable, then you are in good company: Einstein himself unsuccessfully attempted to restore energy conservation to General Relativity. Perhaps energy conservation will reappear within a 'theory of everything' that will (some hope) unite gravity

---

[13] Lecturing physics at a large university, a large part of our job involves correcting the occasional mistruth found in a high-school education in the sciences.

[14] Some textbooks state that the energy lost from light goes into the energy of the expansion, whatever that means. This simply will not do! Universes filled with matter expand faster than those filled with radiation, which is the opposite of what we'd expect if the energy lost by photons 'goes into' the expansion. To close another loophole, the energy lost by photons is *not* balanced by the energy gained in dark energy.

with the other forces, but maybe it won't. For now, we will have to live with it.

If you can move on from this shocking revelation, it's time to consider an interesting question: why are there any symmetries, and hence conservation laws, at all? Remember that the conservation of electric charge is born out of a symmetry of the quantum mechanical wave function. Why does this symmetry exist? In truth, we don't know.

It is worth stressing that there is no problem *in theory* with universes that do not possess the symmetries particular to our Universe. Their equations would perhaps not be as simple or as elegant as those of our Universe, but they are just as *mathematically consistent*. In these different universes, quantities such as electric charge may not be conserved. As we will soon see, enforcing other symmetries can have dramatic results, such as a universe devoid of matter!

So, why do symmetries play an important role in our Universe? Guided by nature, physicists incorporate symmetries into their equations, providing the conservation laws seen in experiments. While these equations mirror nature, they cannot reveal the source of these symmetries. The symmetries in our deepest equations, then, are unexplained ... for the moment.

And yet, in physics, our quest to go deeper into the laws of nature has found *more symmetry* as we go. Deeper problems are often solved with deeper symmetries. So the question of the origin of symmetry is one of the most profound in all of physics.

What if the Universe weren't as symmetric? We'll consider the case of the conservation of electric charge.

On the largest scales, the Universe is controlled by gravity, the weakest of the forces. This might seem like a contradiction, but recall that the strong and weak nuclear forces have a very short range, having essentially no effect outside atomic nuclei. Electromagnetism is a long-range force, like gravity, but opposite charges can cancel out its effect. Wherever positive charges are finely balanced by negative charges, the action of the electromagnetic force is neutralized.

Overall, the Universe is electrically neutral. If the contents of some patch of the Universe were to somehow become predominately positively charged, the corresponding negative charges must be elsewhere in the Universe. The two regions will be drawn together, eventually meeting, mingling, and restoring local electrical neutrality.

It's important to realize just how precisely electric neutrality needs to be enforced. Suppose that you were assembling Earth and got a little careless with your book-keeping, so that for every trillion trillion trillion protons and electrons you put into the mix, *one extra* electron slipped in. The combined repulsion of these extra electrons would be stronger than the attraction of gravity. The Earth would not be gravitationally bound.

In fact, the same net charge (one part in 10^{36}) would preclude any gravitationally bound structure in the Universe at all. Galaxies, stars and planets would all fail to collapse under their own gravity, instead being dispersed by electromagnetic repulsion. The result: a universe of extremely diffuse gas, and not much else.

Our Universe appears to have zero net electric charge, and conserves electric charge, and so this scenario shouldn't keep you up at night. We don't know why the Universe has zero charge, but given charge conservation it seems like a natural state of affairs.

Furthermore, symmetric systems are also more easily analysed and so more predictable. A universe without symmetry, and so without conservation laws, would be chaos. There would be no simple laws to be discovered beneath the complex events around us. Physicist and Nobel Laureate David Gross summarizes this point well:

> Indeed, it is hard to imagine that much progress could have been made in deducing the laws of nature without the existence of certain symmetries. The ability to repeat experiments at different places and at different times is based on the invariance of the laws of nature under space-time translations. Without regularities

embodied in the laws of physics we would be unable to make sense of physical events; without regularities in the laws of nature we would be unable to discover the laws themselves.[15]

Our complaint is not merely that we wouldn't be able to do physics, as if a lack of predictability affects only professors in the laboratory taking their readings. An unpredictable universe is a major problem for an organism trying to survive and a brain trying to organize memories. Life, whether metabolizing, reproducing or processing information, relies upon the stable properties of its own structure and its environment. We're all in trouble if oxygen scrambles its chemical properties tomorrow morning.

So, what is the minimum set of symmetries required for a universe to be capable of forming the structures required for the emergence of complex and intelligent life? For a given set of symmetries, will gas collapse to form stars, and will nuclear reactions be sufficient to warm and energize any planets that condense around them? And will atoms and molecules be stable and versatile enough to support the complex biochemistry of life?

While we can identify some clear disasters, in general we just don't know. The problem is too complicated; we are really only able to scratch the surface. But as our physical theories improve, we may be able to discover the possibilities for life in these other, hypothetical, universes.

But, as we will see in the next few sections, too much symmetry is no good either. A very slight *lack* of symmetry appears to play an important role in why we are all here!

SYMMETRY THROUGH THE LOOKING GLASS

The weak force, with its ability to transform fundamental particles into each other, is different from the other fundamental forces in the Universe. But there is another feature of the weak force that is even

---

[15] Gross (1996).

more mysterious and bizarre, yet often unsung. In the following, it may seem like we have wandered a million miles from the question of fine-tuning, but stick with us. This stuff is really interesting and very important!

Let's start with a story. Imagine you walk into a cinema, where a black-and-white film is playing on the big screen. Being a tale set in Africa, you soar over lush forests and herds of animals running across immense plains. The camera descends on a watering-hole, where a nervous zebra eyes the shadowy shapes of swimming crocodiles. Suddenly, group of people enter the scene, carrying hunting rifles!

A man steps forward and raises his rifle to shoot.[16] Something doesn't look right, as he grips the barrel with his right hand and operates the trigger with his left. The shooter is left-handed. This isn't particularly unusual; roughly one in ten people need those 'special' scissors. But your intrigue grows as the other shooters begin firing, and they too all hold their rifles left-handed. So many left-handed people meeting together seems quite strange.

As you look closer, something else seems amiss. Their guns, bolt-action rifles, would normally have the bolt on the right side, so that it can be worked with the right hand, but these hunters have modified rifles with bolts on the left-hand side. With these modified rifles, the hunters are able to efficiently use their guns, rather than have to clumsily reach across the gun to reach a right-handed bolt. Curious.

But, as you watch, you notice something very strange. The film has zoomed out to reveal the hunters' truck, and that truck has words written along the side. And while you can clearly read the words, 'The Hunters Club', you see that the lettering has been flipped:

---

[16] No animals were harmed during the writing of this book. All testing was done on graduate students.

The Hunters' Club

You realize that the film you have been watching is not a tale about a group of left-handed hunters with specially built rifles, but is actually about right-handed hunters with off-the-shelf weapons. And the projectionist has flipped the film when putting it into the projector.

Why didn't you notice earlier? Through all of those majestic scenes of jungles, plains, and animals at the watering-hole, not once did you realize that the film had been flipped.

It was only when something asymmetric came onto the scene, such as the right-handed dominance in humans, that you suspected that something was not quite correct.[17] However, this is not a strong asymmetry as left-handedness is not that rare. Seeing a group of left-handers, while interesting, is not impossible. Similarly with the left-handed rifles, as some specifically left-handed equipment and implements exist.[18]

However, the orientations of the letters were a dead give-away. Letters and words are unmistakably asymmetric, and clearly the world you were watching on the film was not the world you inhabit.

Asymmetry in nature goes deeper than the use of our limbs. Many of the molecules inside living bodies are asymmetric, and this affects how they interact with other molecules. Take, for example,

---

[17] An actual mirror-image version of the Earth was central to the classic 1969 science fiction film *Doppelganger*, also known as *Journey to the Far Side of the Sun*. The fact that it is a mirror-image planet is slowly revealed through noticing unexpected asymmetries.

[18] Geraint is right-handed, but eats in a left-handed configuration. He is often thwarted at restaurants when requesting a left-handed fish knife. While left-handed implements exist, they appear to be rarer than left-handed people. Luke, also right-handed, attempted to use left-handed scissors in 2004 and is still recovering.

R-CARVONE   S-CARVONE

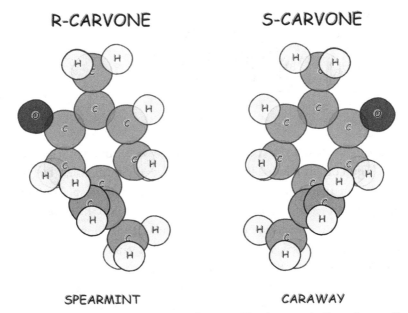

SPEARMINT   CARAWAY

FIGURE 33 Two isomers of carvone. To a human, the R-version smells of
spearmint, while the S-version smells of caraway, showing that scent
receptors must be able to tell them apart and hence they too are asymmetric.

carvone. This naturally occurring molecule is quite complex, with 25
atoms arranged as shown in Figure 33. There are, in fact, two ways to
arrange carvone, with one being the mirror image of the other.
Chemically, they seem to be the same molecule, until you inhale
a sample. The 'right-handed' version smells like spearmint, whereas
the mirror-imaged molecule has the scent of caraway seeds.

Similarly, limonene has an orange-smelling form, and a mirror-
image that smells of lemon. Certain sugars and carbohydrates that are
absorbed by your body during digestion have mirror molecules that are
useless, and are simply excreted.

Your body's preference for molecules of a certain handedness
runs deep. The amino acids from which your cells make proteins,
tissues and organs are all left-handed. A single right-handed amino
acid in a protein would cause it to fold differently, creating a very

different (and potentially harmful) molecule. Life's left-handedness is still something of a mystery; some scientists think that the weak force is part of the answer.

How deep do these asymmetries go? Are there asymmetries written into the laws of the Universe?

The Universe, as we've seen, is governed by fundamental particles and fundamental forces. Let's imagine we walk into our cinema and find a film of an electron interacting with a photon, or quarks interacting with gluons within the nucleus of an atom. Now, as with the hunters, we ask if we can tell the difference between the original film and one that has been mirror-flipped.

This seems unlikely. Electrons and quarks do not carry little letters on them, and neither do photons and gluons. Nothing would seem amiss in a mirror-imaged film. An electron enters from stage-left while a photon comes from stage-right, they bounce off each other and leave the way they came; this scenario is equally possible with the entrance roles reversed. The fundamental laws appear to be left–right symmetric. In physics parlance, the laws are *parity invariant*.

Once again, the weak force is the odd one out. The weak force violates this parity invariance! We will explain what this actually means in a moment, but when it was discovered in the mid 1950s, it came as a shock. It seemed obvious to almost everyone that all interactions 'conserved parity', in that the mirror-flipped scenario was just as possible as the original. How could the Universe know the difference between left and right?

But it does. Understanding the weak force has been a difficult process, but through the 1930s and 1940s physicists started to figure out what was going on when, for example, electrons and positrons are spat out of the nuclei of atoms. Some began to wonder whether the weak force obeyed parity symmetry. Experiments to test this were proposed. These are probably the most important experiments that you have never heard of!

One particular experiment was proposed by two young researchers in the USA, Chen-Ning Yang and Tsung-Dao Lee, and performed

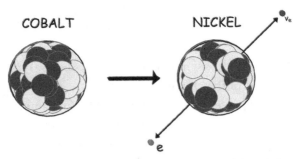

FIGURE 34 The Wu experiment. The decay of the cobalt nucleus is strongly asymmetric, with electrons only being emitted downwards with regards to the spin of the nucleus. This is because the other decay particle, the neutrino, can only spin one way. How does it know about left and right in the Universe?

by Chien-Shiung Wu at the Bureau of Standards in Washington, DC. This delicate experiment examined the radioactive decay of the element cobalt.

Cobalt is a rather heavy nucleus, with 27 protons and 33 neutrons, and, with a radioactive half-life of just over 5 years, it decays into nickel through the weak force, spitting out an electron in one direction, and a neutrino in the other (Figure 34). Due to its fleeting nature, the neutrino escapes without detection, but the electron can be easily captured.

A scientist with an electron detector near a lump of cobalt will see electrons being spat out in all directions, with a similar flow of invisible neutrinos away from the lump. It all appears nice and symmetric, so where is the asymmetry sought by Yang and Lee?

To see this, we need to convince the cobalt nuclei to line up. But this might seem strange, as nuclei are balls of protons and neutrons. What do you use to orient the nuclei if they are to line-up?

As we noted earlier, the fundamental particles, the leptons and the quarks, possess spin. Similarly, nuclei spin, as they are composites of particles. Each cobalt nucleus acts like a little magnet, with the direction of the spin acting like a compass needle. If we place our cobalt nuclei in an external magnetic field, their individual spins all point north.

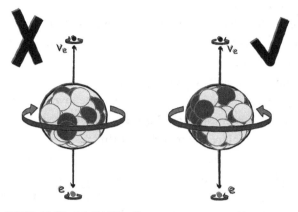

MIRRORED UNIVERSE   OUR UNIVERSE

FIGURE 35 In our Universe, in the decay of cobalt into nickel, electrons
are emitted in the direction opposite to the spin (viewed as anticlockwise
from above; right-hand). In the mirror view of the Universe (left) the
electron is emitted in the same direction as the spin, something that is not
actually observed in our Universe.

So, here's the experiment. Place your cobalt nuclei in a strong
magnetic field, with the spin pointing up. Some of these nuclei will
decay. Like iron filings lining up in the presence of a magnet, the
emitted electrons will want to move along the field lines, but being
emitted up or down should be equally likely. However, this is not what
is seen. Electrons are only emitted down, *against* the cobalt spin.

So what, you might think. But imagine holding a mirror up to
the decay of the cobalt nucleus; what do you see? In the mirror, the
spins of the nucleus and electrons reverse. Looking down from the top,
they are spinning clockwise instead of anticlockwise. However, top
and bottom remain the same. In this mirror universe, electrons are
emitted in the *same direction* as the cobalt spin, not against it. This is
never seen in our Universe (Figure 35).

Think about how bizarre this is for a moment. Through all of the
physics you could see on a film, anything involving the strong force,
electromagnetic force, or gravity, none would reveal to you whether
your film has been flipped. But if you were shown a film of cobalt

nuclei decaying in a magnetic field you could examine the spin and the direction that electrons are emitted, and immediately tell whether you were looking at our Universe, or a mirrored version. The universe in the mirror works differently!

So astounding was this result, Yang and Lee won one of the fastest Nobel Prizes in history;[19] Feynman, meanwhile, lost a $50 bet with fellow scientist Norman Ramsey. Scientists still struggle to understand and appreciate the meaning of this result.

Why is the weak force asymmetric, when all of the other forces are symmetric? It comes down to that elusive particle, the neutrino. To understand this, let's look at one of the neutrino's more popular cousins, the photon. Like other particles, photons spin. If we could look at the spin of photons travelling towards us, we'd see some spinning clockwise while some would be spinning anticlockwise. Putting a mirror up to a photon spinning clockwise produces a photon spinning anticlockwise, something with which our Universe is quite OK.

However, neutrinos refuse to conform to this symmetric stereotype. If you examined the spin of neutrinos travelling towards you, you would find them all spinning in the clockwise direction. All of them![20] They know about the left and right of the Universe!

Why does the Universe behave this way? The truth is nobody knows, but as we will see next, these kinds of subtle asymmetries play a very important role in why we exist at all.

MORE MATTER THAN ANTIMATTER

As we've mentioned before, the particles of the Standard Model are accompanied in this Universe by antiparticles, such as the electron with the positron. Antiparticles are just like regular particles except that all of their quantum numbers are flipped. For example, the electron and positron have opposite charges.

---

[19] Yang and Lee's paper predicting parity violation was published in 1955, and the pair took the Nobel Prize in 1956. By way of comparison, Einstein received the 1921 Nobel Prize for his contributions to physics, especially those of his 'miraculous year', 1905. Wu, whose experimental work proved the existence of parity violation, was not a recipient of the prize!

[20] There is a subtlety here. While neutrinos all spin in a clockwise direction, their antiparticle, the antineutrinos, all spin in an anticlockwise direction.

And just like electrons, quarks have antiquarks, and baryons composed of antiquarks make antibaryons. So we can make antiprotons and antineutrons. With antiprotons and positrons, we can build anti-hydrogen. In fact, we could construct the entire antiperiodic table, chemically identical to its matter counterpart. Antihydrogen will emit the same spectrum of light as hydrogen, and anti-hydrogen will happily join with anti-oxygen to make antiwater, which will flow and freeze and splash in exactly the same way as normal water.

Indeed, we could construct an anti-human who would, for all intents and purposes, be identical to you and me. However, if you ever meet such an anti-person, don't shake hands! Your protons, neutrons and electrons and their antiprotons, antineutrons and positrons would annihilate, producing an intense flash of photons.

But one significant difference between matter and antimatter in our Universe is that, while matter is common, antimatter is extremely rare. Why?

To understand this, we have to go back to the very earliest moments of the Universe. As we saw in the previous chapter, these initial moments were marked by an extreme accelerated expansion known as *inflation*.

Inflation is required to iron out some of the peculiarities of the standard cosmological model, but it is still poorly understood. The mechanism that started it, drove it, and eventually ended it is still the subject of ongoing debate. Also, the duration of inflation is critical to the life-bearing properties of the cosmos, in particular the seeding of cosmic structure.

Inflation plays another important role in our existence. As inflation proceeds, all the matter and energy in the Universe are thinned out to an unimaginable degree, replaced with the energy of the inflaton (the *whatever-drives-inflation* field). When inflation ends, this energy is converted back into matter and radiation in a process known as *reheating*. Again, the process of reheating is poorly understood. (Are you starting to note a theme here?)

In this seething sea, photons collide to create particles and antiparticles, and particle/antiparticle pairs annihilate into photons. This constant creation and destruction of particles and radiation is governed by the conservation laws, so while the number of electrons and quarks might fluctuate, the net charge in the Universe remains zero.

But, what of the other conservation laws? Baryon number, that is, the number of baryons minus the number of antibaryons, appears to be a conserved quantity in our Standard Model of particle physics.[21] Thus, while the number of protons and antiprotons fluctuates, the net baryon number should remain zero.

As the Universe expands and cools, the energy of each photon in the sea of radiation decreases. To create a particle/antiparticle pair, you must supply at least enough energy to make their *mass*. Eventually, the photons in the Universe have too little energy to form neutrons and antineutrons, protons and antiprotons. A short while later, even the less-massive electrons and positrons are too much to ask. These particle/antiparticle pairs continue to annihilate into radiation. Eventually, all of the particles and antiparticles vanish into the radiation sea. And because of the conservation of baryon number, we would expect effectively zero protons and neutrons in the universe.

Hmmm. We've followed physics as we know it, and reached the wrong conclusion. There are, you may have noticed, plenty of protons, neutrons and electrons around. You're made of them. So, why are there protons and neutrons still available in the Universe to create stars and planets, trees and people? Why is there more matter than antimatter? We'd better reconsider our scenario above and try to spot where we went wrong.

---

[21] For the experts, processes involving *sphalerons* can violate baryon and lepton number, while maintaining their difference. However, 'while the Standard Model satisfies all of the conditions for baryogenesis [including baryon number violation], nothing like the required baryon number can be produced' (Dine and Kusenko, 2003). We will simplify our discussion by focusing on the asymmetries *beyond* the Standard Model that are required.

For the past few decades, cosmologists have tried to unravel this mystery. To date, we don't have a complete, tested, and successful theory from particle physics to explain why we are made of matter and where all the antimatter went.

But there are clues. We can estimate how much asymmetry we need by measuring the number of baryons in the Universe and comparing with the number of photons in the cosmic microwave background; this radiation is left over from the annihilation of particles and antiparticles in the early Universe. Our Universe contains roughly one particle for every billion photons. Thus, for every billion antiprotons in the very early Universe, there were a billion *and one* protons.

The great Russian physicist Andrei Sakharov identified some of the *necessary* conditions for creating more matter than antimatter, conditions that are missing from the Standard Model of particle physics. One non-negotiable condition is that baryon number conservation has to go. A subtle kink in the mathematics of particle physics must cause the total number of baryons (minus antibaryons) to not be perfectly conserved.

Whatever the cause of this asymmetry, it is clear that life needs matter! If there were no matter/antimatter asymmetry, our Universe would be nothing but a cooling sea of radiation, completely devoid of the particles that make nuclei and atoms. This is a wonderfully clear case of a life-prohibiting universe – there's no structure at all! Sure, radiation is nice, but doesn't do much on its own.

But why is the asymmetry one part in a billion? Why is it not one part in a trillion? Or one part in two? To answer this, we need to know the source of the asymmetry. The new asymmetric force we need could be the *X-force*, which we encountered in Chapter 4. You'll remember that the X-force can turn baryons into leptons, violating baryon number conservation. As the number of baryons was not conserved in the earliest instances of the Universe, so it will not be conserved in the far future, due to proton decay. The very process that brought our atoms into being could ultimately result in them melting back into radiation.

208 ALL BETS ARE OFF!

In this case, the amount of asymmetry has something to do with the strength of the X-force. If the force were too weak, then there would be too little asymmetry and so not enough matter in the Universe to make anything interesting at all. On the other hand, if the X-force were too strong, protons would decay quickly. In the 1940s, physicist Eugene Wigner understood that the lifetime must be long; as he said, 'I can feel it in my bones'. With more rapid proton decay, we would glow as the protons that make our very atoms decay away. Even if protons lasted ~10^{14} years, the energy dumped into our bodies would result in genetic mutations and cancer.

While the strength of the X-force looks like a case of fine-tuning, we should reserve judgement. We don't know whether the X-force is responsible for matter/antimatter asymmetry, and we don't know enough about its properties to know whether they are fine-tuned. In the cases of fine-tuning discussed in previous chapters, such as the stability of nuclei or the formation of galaxies, we are dealing with familiar physics, successful theories whose mathematical properties are well understood and whose predictions have been thoroughly tested against experiment. These cases of fine-tuning are grounded in what we *know*, not in our ignorance. Unfortunately, we just don't have that for the physics of matter/antimatter asymmetry. We *suspect* fine-tuning, but in the absence of a well-understood, well-tested theory, we can't demonstrate it.

Nevertheless, we are left with nagging questions. Why does our Universe allow a mild amount of baryon asymmetry, large enough to produce the matter that makes you and me, but small enough not to dissolve our protons? Why does our Universe *not* allow similar asymmetry with regards to charge conservation, the existence of which would lead to such chaos?

UNHAPPY COUPLINGS!

There is another aspect of forces that we need to consider. Within our Universe, the fundamental forces are neatly divided: mass talks to mass via gravitation, and electric charge talks to electric charge via electromagnetism.[22] But within the mathematical framework of quantum mechanics, we can describe other ways for particles to talk to each other.

We have to clarify what we mean by 'talk'. Take, for example, the electron. In quantum mechanics, an electron is a vibration in the *electron field*. To 'talk' amongst themselves, electrons generate vibrations in the electromagnetic field, which are photons. So when we say that an electron talks to another electron, we mean that one creates a vibration in the electromagnetic field that is felt by the other. How the electron field interacts with the electromagnetic field is governed by the coupling constants discussed earlier.

The Standard Model of particle physics presents all the ways that particles (and their fields) talk to each other in our Universe. But the mathematics of quantum field theory, which underlies the Standard Model, permits many other ways that these particles could chatter amongst themselves. This would imply the existence of other fields, other coupling constants, and hence other forces to move and transform particles.

Such forces can be added to our equations, and the predictions of such a theory can be tested. Thus far, experiments have failed to detect any such couplings in our Universe. The Standard Model is, in this sense, surprisingly clean and uncluttered.

Is a lean Standard Model an ingredient for complex life, or does it just make the Universe easier to understand? And if it were more bloated, what then would be the effect of these other possible forces on life in a universe? Could an additional interaction between electrons and quarks mean that stable atoms could never form? While we can imagine such scenarios, the mathematical possibilities are too numerous to fully explore. So, as of yet, it is difficult to judge whether the

---

[22] If electromagnetism combines electricity and magnetism, then why do we only talk about electric charge? What about magnetic charge? The equations of electromagnetism have a space waiting for magnetic charges, and in fact the equations would be more symmetric if they were included. So far, no trace of magnetic charge (known as a *magnetic monopole*) has been found. All magnetism in our Universe is the result of electric charges moving or spinning.

cleanliness of the Standard Model is necessary for the formation of complex and intelligent life in the Universe. But in considering the possible state the Universe could have found itself in, we would certainly have to think about it.

ON THE NATURE OF TIME

The nature of time has puzzled many people for a long ... uh ... time.

> For what is time? Who can easily and briefly explain it? Who even in thought can comprehend it, even to the pronouncing of a word concerning it? But what in speaking do we refer to more familiarly and knowingly than time? And certainly we understand when we speak of it; we understand also when we hear it spoken of by another. What, then, is time? If no one ask of me, I know; if I wish to explain to him who asks, I know not.
>
> The Confessions of St Augustine (AD 415)

Think of the features of life that we have discussed thus far – metabolism, reproduction, information processing, thinking, book writing. These all *take time*. They are *processes*. And so we have to ask, what is time?

Perhaps our most basic impression of time is that it *passes*. The current moment soon becomes the past, replaced by something that was previously the future. However, it seems that we are just defining time in terms of other *temporal* words: *moment, becomes, past, previously, future*. This makes talking about time a bit tricky. The idea that time *passes*, for example, would suggest that time itself is moving, which doesn't make much sense.

Luke's Special Relativity lecturer, Professor Tim Bedding, had an easy answer: time is what you measure with a clock.[23] For the physicist, this seemingly trivial definition is enough to get us started. It is an *operational* definition: we can study physical processes and compare the rates at which they occur.

---

[23] Luke's *General* Relativity lecturer was Geraint. Just so you know who to blame.

However, there is more to time than its *intervals* measured by a clock. There is an *arrow of time*. The sequence of physical states of the Universe is profoundly asymmetric, as just about any video played in reverse will demonstrate. A dropped glass shatters, but does not spontaneously repair itself. Life uses this asymmetry to function, reproduce, and think. For example, we form memories of the past, but not of the future. Where does this arrow of time originate?

Radiation Arrow of Time

Let's begin with the familiar. Our sense of sight is amazing. We rarely give it a second thought, but optical light is continually flowing into your eyes, through the dark pupils at the centre of your colourful irises, and onto light-sensitive cells that translate the image into electronic signals to be sent to your brain.

All detectors must find a compromise between competing features – design committees for new telescopes move painfully slowly – and your eyes are no exception. Because of the ways its sensors are wired, there is a gap in the retina that is devoid of light-detecting cells. This *blind spot* was discovered by Edme Mariotte, and you can experience it for yourself via a simple experiment – Google it!

No one noticed this blind spot until the gap in our retina was discovered, and so your brain does an excellent job of taking incomplete information and presenting your consciousness with a continuous, high-fidelity picture of the world around you, reconstructed from trillions and trillions of photons in a narrow range of frequencies. In fact, some of our cousins in the animal kingdom can see into the infrared and ultraviolet, revealing a world that we cannot see without specialized equipment.

Accompanying the photons of various wavelengths, Earth is bathed in particles – electrons, protons and neutrinos – raining down from the Sun and distant cosmos. In fact, if your eyes were sensitive to neutrinos, you'd be able to see straight through the Earth and into the very heart of the Sun! Alas, a portable and effective neutrino 'eye' cannot be made from any known material.

The *light* arriving from the Sun is predominantly in the optical range of wavelengths, which can interact strongly with atoms and molecules. Our eyes use this bonanza of electromagnetic radiation. Because of the finite speed of light, when we look into the sky we see an image of the Sun as it was eight minutes ago. And as we look deeper into space, we are seeing the nearest star, Alpha Centauri, as it was more than four years ago. The light we receive from the Andromeda galaxy started its journey over two million years ago, long before our ancestors mastered this fire business. The light we see in the cosmic microwave background has been travelling for almost 13.8 billion years.

We can understand all of this using the equations of electromagnetism. And yet, these equations have a curious feature. Take any scenario permitted by those equations – such as light emitted by a lamp, bouncing off a painting and into your eye – and run it *backwards*. This reversed scenario is perfectly OK, according to the equations. Both are possible.

And yet, the backwards version of seeing a blue sky on a sunny day is rather peculiar. Signals from your brain excite your retina, causing photons to be emitted from your eyes, which travel upwards to scatter off the upper atmosphere, and then cruise through space towards the Sun. They are joined by electromagnetic waves from all over the Universe, converging towards the Sun. (Think of a stone thrown into a pond, but in reverse: the circular ripples all converge back towards their centre, until a stone pops out of the water!)

We don't see these converging waves, but what would they look like? This is going to sound like something from science fiction, but a converging electromagnetic wave could be equally well described as an electromagnetic wave from the future! If this is how electromagnetic waves work, we must wonder why we observe an image of the past, but do not similarly see a vision of the future.

What would light from the future look like? Suppose that a star from the future wandered into our sky. To us, it would not shine brightly, pouring radiation out into the Universe. Rather, it would be a colossal light collector, accumulating photons sent in its direction from all parts of the Universe. Light emitted from your eyes would join a converging chorus of radiation, destined to enter the star, scatter through its hot gas and ultimately use its energy to split helium into hydrogen.

The human mind is a complicated thing, so it's difficult to know exactly what it would be like to experience light from the future. Consider, instead, a video camera pointed at the dark star. A recording of the light 'from' the star would be steadily erased from the tape, converted into light and sent out of the lens of the camera. Thinking a bit about a human mind, this would suggest that you would have a memory of the future in your head, which is forgotten when your eyes emit these future photons. Take a moment to think: is there something you are about to forget, like next Saturday's lottery results?

By now you are probably thinking this is all very interesting, but it is obvious why we don't receive light from the future, because that is silly. But, as we mentioned, there is nothing in our laws of physics that prevents this. For the laws to account for our experience of the Universe, we must understand why the mathematical possibility of light from the future doesn't actually happen in our Universe.

Matter, too, has a complicated relationship with time. Within the Standard Model of particle physics, an electron sent backwards in time looks like its antimatter sibling, the positron! In fact, all antimatter particles can be viewed as their matter twins travelling backwards in time.

This might seem like a quaint mathematical oddity, but for a brilliant insight by John Archibald Wheeler, one of the giants of twentieth-century physics. His most famous student, Richard Feynman, told this story in his Nobel Prize lecture.[24]

> I received a telephone call one day at the graduate college at
> Princeton from Professor Wheeler, in which he said, 'Feynman,

[24] Available at www.nobelprize.org.

I know why all electrons have the same charge and the same mass.'
'Why?' 'Because, they are all the same electron!'

According to Wheeler's idea, there is only one electron in the
Universe, furiously travelling backwards and forwards in time. All of
the electrons and positrons we see around us are a snapshot of the
same electron passing through the present moment in time. While
ingenious, Wheeler's idea predicts that the number of electrons in our
Universe will be the same as the number of positrons. As we saw
above, this is not the case.

Nevertheless, Wheeler's idea illustrates the *time-reversal symmetry* of the Standard Model of particle physics. Well, almost. Just as
the weak force can break parity symmetry, it can break time-reversal
symmetry, but very rarely and weakly. This very slight asymmetry in
our laws does not explain why the state of our Universe is so profoundly asymmetric with respect to time.

Perhaps we need to extend our search to the Universe as a whole.

Cosmological Time

According to the Standard Model of Cosmology, time (as we know it)
was born in the Big Bang approximately 13.8 billion years ago, and
marches on into the infinite future. But if time is measured by a clock,
what clock?

In Einstein's General Theory of Relativity, space and time are
mere coordinates. The particular set of numbers that we assign to
a place and time in the Universe is a matter of convenience. Think
of coordinates on a map of the Earth – it's not as if rocks and mountains come labelled with latitude and longitude. We could change the
labelling and still be able to navigate.

And yet, the choice of latitude and longitude isn't totally arbitrary. The Earth is rotating, and roughly spherical. It has a centre,
which coincides with the centre of each circular line of longitude.
The point where the lines of longitude meet is where the Earth's
rotation axis touches its surface – the South and North poles. A slice

through the Earth at constant latitude meets the rotation axis at right angles. While the Earth doesn't come with labels, it does come with *symmetry*, both in its geometry (spherical) and its motion (rotation). Our coordinates can *reflect* this symmetry.

As we noted in Chapter 5, the Universe is also highly symmetric. The Universe is the same everywhere (*homogeneity*) and looks the same in all directions (*isotropy*). There is a time coordinate that reflects this symmetry, known as *cosmic time*. If you look at all parts of the Universe at the same moment of cosmic time, you'll find that they all have (on average) the same matter density, and the same cosmic microwave background temperature.

Clocks that travel along with the expansion of the Universe measure cosmic time. In this way, we can meaningfully state that the Universe is 13.8 billion years old.

Does the expansion of the Universe with respect to cosmic time account for our experience of time? Not quite, and for reasons that will sound familiar.

The problem is that the equations that describe the expansion of the Universe will also allow us to consider time running backwards. If you take a solution of those equations and reverse time, you get an equally valid solution. Our Universe, with its Big Bang in the past and eternal expansion in the future, has a time-reversed twin with an infinitely contracting past and a future big crunch.

So, the expansion of the Universe exhibits an *arrow of time*, in that the Universe was very different in the past and will be very different in the future. But the equations that describe the expansion of space don't explain why we see a Big Bang in our past but not in our future. The changing scale of the Universe doesn't explain the direction of the arrow of time.

A Quantum Arrow of Time

If we take Einstein's relativity at face value, the Universe looks like a map that is already drawn. Spacetime just is, and our journey through spacetime is like a path on a map. While we are only part way on our unexpected journey, the Misty Mountains and Brandywine River are 'already' there, waiting for us, so to speak. In this view, there is right now a fact of the matter about what will happen tomorrow.

Our experience of the world, however, is that the future is open, free to be influenced in one way or another by our actions. How can our choices impact the future if all we are doing is following a path through pre-existing territory? The solution, some propose, is that the future map has not yet been drawn and it is constantly unfolding in front of us. The past is fixed and knowable, in contrast to the future, which is open and unpredictable. In such a picture, it's obvious why we don't get photons and matter from the future, as the future does not yet exist!

The *indeterminacy* of quantum mechanics is relevant here. In some interpretations of the theory, the state of the world at any particular time does not determine its state at some future time. The radioactive nucleus that we create in the lab today might decay tomorrow, or the day after – and there is nothing about the laws, or the nucleus, or even the whole Universe that says when the decay will happen. Today, there is simply no fact about whether the nucleus will decay tomorrow. There are probabilities, but the future is a story yet unwritten.

The problem is that the most popular non-deterministic inter-pretations of quantum mechanics are – as physical theories – incom-plete. The part of the interpretation that is time asymmetric is where a *measurement* is made, which forces the quantum system to choose between its possible (superposed) states. But this places physicists and their apparatus above the rules of quantum mechanics, leaving a gaping hole in the theory.

Putting the issue of measurement aside, the central law of quantum mechanics, Schrödinger's equation, is time-reversal sym-metric. As we noted above, the fundamental, quantum laws of matter and radiation show no sign of the time asymmetry we see in the Universe at large. If there is no arrow of time in the laws that govern the basic stuff of the Universe, then where is it?

Thermodynamic Time

There are no fluids in the basic stuff of the Universe, either. The wind blowing through the trees is the result of trillions of trillions of particles working together to flow, compress, expand, swirl, wave and more. Fluid properties *emerge* when lots of particles work together. Perhaps the arrow of time is also an *emergent* phenomenon.

In particular, we have already noted the important connection between thermodynamics and time. If you knock a coffee cup off a table, it falls onto the floor and smashes into pieces, spilling its delicious contents. If you want a hot beverage, on the other hand, it is a bad idea to spread shards of broken cup and tepid coffee on the floor and wait for them to reassemble, heat up, and leap back onto the table. You've never seen *that* happen in real life, and yet it is the time-reversed version of the cup breaking.

All fairly obvious, and yet the underlying equations of physics do not contain an arrow of time! The magical reassembling coffee cup fits as neatly into our equations as every accidental breakage you've ever seen. Even after the cup has smashed and splotched its contents everywhere, if we could stop time and reverse the directions of all the particles' velocities, we would see the coffee unsplosh its way back into the cup and onto the table. In particular, sound waves in the surrounding air would converge on the reassembling shards at just the right place and with just the right energy to fuse the shattered pieces back together.

Here we start to see an asymmetry. Just about any old push will result in the cup smashing on the floor, whereas reassembling the cup requires an impossibly precise coordination of every atom and molecule of cup and coffee, air and floor.

If you've been paying attention, some of this sounds an awful lot like our discussion of entropy in Chapter 4, where things colloquially go from order to disorder. And you'd be right. The thermodynamic arrow of time points in the direction of increasing entropy.

We have, then, what looks like a neat explanation for the arrow of time. Any old fall breaks the cup, but only an extremely unlikely set of coincidences reassembles the cup. So we tend to see, as time goes by, breaking cups but not spontaneously assembling ones. We can make a similar argument for other entropy-increasing processes: any old set of thermal collisions between molecules will melt the ice in your drink, but it would take a very special (and so unlikely) set of circumstances for your drink to turn back into lukewarm water and ice. Stars are entropy producers, so the star 'from the future' would require a similarly unlikely set of coincidences. And quantum decoherence, which we met earlier in this chapter, shows how quantum measurement could be linked to the increase of entropy as a system interacts with its large, messy environment.

There are, however, a few loose ends. Firstly, why does your personal arrow of time point towards the direction of increasing entropy? That is, why do we remember the low-entropy past and not the high-entropy future?

Important clues come from thinking about computers, because running a computer program increases entropy. In particular, storing information in the memory of a computer is *irreversible*[25] since the computer overwrites and erases information. (As you may have guessed, this document was saved and backed up as soon as this sentence was written.)

So, if the brain's physical memory works like a computer in this way, then it can only form memories of the past, that is, of the time when the Universe's entropy was lower. This, then, is good reason to believe that any information-processing life form will experience the

---

[25] This result is related to *Landauer's principle*. The idea is that a computer program transforms an initial state into a unique final state: given 4 and 2, a program can be told to add these numbers, giving 6. But seeing only the result, we have no way of uniquely inferring the numbers that were added. This information has been erased from the computer's memory, and the operation is irreversible. Any physical implementation of the program must create an increase in entropy. Simply put, computers get hot because they have to forget.

thermodynamic arrow of time, always remembering the thermodynamic past but not the future.

The second loose end is more worrisome. We seemed to have uncovered a thermodynamic arrow of time, but where did the time-reversal symmetry of the fundamental laws go? How did we pick out a special time direction when the laws themselves don't?

The problem is: *we didn't*. Suppose that at 10 a.m. we made ourselves a glass of water with ice cubes. At 10:15, the cubes have partially melted, and by 10:30 we have only a tepid glass of water. Let's look closely at the glass and ice at 10:15. By thinking about order and entropy, we correctly guess that the most likely thing to happen next is that the ice will continue to melt, until eventually all the water has reached room temperature.

The problem is that we reach *the same conclusion* about the glass at 10 a.m., because the fundamental laws are time-reversal symmetric. Starting at 10:15 and predicting backwards, the most likely thing that came at 10 a.m., *before* the half-melted ice, is also room-temperature glass. The logic is impeccable – low-entropy states are more unlikely. So, to explain the half-melted ice in terms of unmelted ice is to explain the unlikely by appealing to the even more unlikely.

Wait, you think, I remember the unmelted ice at 10 a.m. Even this doesn't matter: your memory is another physical state, and so should not be explained by such an incredible extravagance as an even lower entropy state.

Thus, time-reversal symmetry returns with a vengeance! Given any low-entropy present state, and according to the same reasoning that led us to correctly predict that the *future* will see increasing disorder, we would predict that the *past* is similarly disorderly. This is in violent conflict with all our memories, including our records of all the experiments from which we learned the laws of thermodynamics!

The physicist Niels Bohr was supposedly fond of joking that it is difficult to make predictions, especially about the future. When it comes to a statistical understanding of thermodynamics, exactly the opposite is true. Our predictions of the future are quite successful – but our 'predictions' of the past are an abysmal failure.

Something's gotta give. Remember the discussion of free energy in Chapter 4; the increase in entropy of our Universe points back to its extremely low-entropy beginning. Our Universe was born with abundant free energy, with the potential to undertake the many processes that eventually led to us.

This suggests a way out of our conundrum, a way known as the *past hypothesis*. We simply postulate that the Universe began in a low-entropy state, and that all our reasoning about what states are likely and unlikely must start from this assumption. Instead of finding the arrow of time in the laws of nature themselves, we find it in the Universe's very special initial state. Entropy will increase with time in one direction because such processes are the most likely. Entropy will not similarly increase in the other time direction because there is no going backwards from the *beginning*.

Isn't this just cheating? After all, low-entropy states are supposed to be unlikely, and so simply positing that the Universe began with very low entropy seems to be yet another problem, not a solution.

But think more carefully about how we came to conclude that low-entropy states are unlikely. To turn a lower-entropy 'intact cup on a table' arrangement into a higher-entropy 'smashed cup on a floor' arrangement required a generic shove, while the reverse transition required a practically impossible conspiracy of delicate molecular motion. So what we should really say is: low-entropy states are unlikely to result from high-entropy states.

It follows that the past hypothesis isn't unlikely *on those grounds*, since the initial low-entropy state does not result from a higher-entropy state. The past hypothesis must be taken as given before we make any judgements about what states are likely and unlikely.

Nevertheless, all of this leaves us with an uneasy conclusion. From what we have learned, the passage of time, how we perceive it and how we experience it, appears to be a consequence of the very special initial state of the Universe. Why the Universe was born in such a low-entropy state, when it could have been born with all available energy locked up in high-entropy black holes, is a mystery. A high-entropy beginning would leave little or no free energy to power metabolism, reproduction, information processing, memory forming, or any of the activities we would call life. Our discussion here shows the depth of this enigma, entwined as it is with the nature of time itself.[26]

DIMENSIONS: ONE, TWO, MANY?

Space, like time, is one of those things that we often take for granted. We have three dimensions in which to live and travel, and one time dimension, ticking away the hours of our busy days. As expected, we're going to ask: what's so special about *three* (for space) and *one* (for time)?

Would there be life in the universe if it contained only two dimensions of space, or maybe even one? What if we considered the other extreme where we had five hundred dimensions of space?

While thinking about the difference between a two-dimensional sheet of paper and a three-dimensional block of wood might help us comprehend a couple of additional dimensions of space, what about extra dimensions of time? This sounds like science fiction, but mathematically it's easy to add a couple of extra time dimensions into physics. I'm sure all of us would like a couple of extra time dimensions we could pop into when a deadline is due, sliding back into the original with just enough time to impress the boss. But would such universes be convivial to life?

---

[26] David Albert's *Time and Chance* (2004) is an influential and accessible defence of the past hypothesis. See also Sean Carroll's *From Eternity to Here* (2010), and Winsberg (2012). Of particular interest to defenders of the past hypothesis should be the rapid rise in the physical sciences of Bayesian probability theory, which views probability as quantified uncertainty. We will discuss Bayesianism further in Chapter 7, but note that this would provide a probability measure over and above even the past hypothesis.

We can recast the laws of physics into a form that can handle differing numbers of dimensions. And the results aren't good. Strange things happen when you start to play with the number of dimensions, things that you would not necessarily expect.

Let's take a simple example: Newton's law of gravity. The gravitational force is responsible for keeping the Earth, the other planets, comets, and asteroids in their orderly dance around the Sun. Newton gave us the precise mathematical description of this force between two objects, being proportional to the product of the masses and inversely proportional to the square of the distance between them.

In our three-dimensional universe, the gravitational force falls off as the square of the distance, the force diminishing in strength as it spreads out over larger and larger areas, in the same way that a light appears dimmer as it is moved to greater distance.

We can write Newton's equation in a way that lets us change the number of space dimensions.[27] Let's now increase the number of spatial dimensions from three to four. With one more dimension, the strength of the gravitational force would decrease as the inverse *cube* of the distance. The result? Chaos!

Planetary orbits in such a universe could not be *roughly* circular. They would have to be *exactly* circular, as the slightest deviation from perfection or perturbation by a nearby planet would send the Earth spiralling into the Sun or racing away into empty space. We would fry or freeze.

A similar fate would await orbiting electrons in atoms. In a universe with more spatial dimensions, there is no *ground state*. That is, there is no lowest energy orbit below which the electron cannot fall. Electrons would spiral into nuclei; farewell, atoms and chemistry!

We show in Figure 36 the possibilities for different numbers of space and time dimensions. 'We are here' at 3 space + 1 time dimension, and universes to our right – with more space dimensions – are

---

[27]   That is, Poisson's equation. Or, for bonus marks, use Einstein's equations of General Relativity to derive the Schwarzschild metric in n-dimensions.

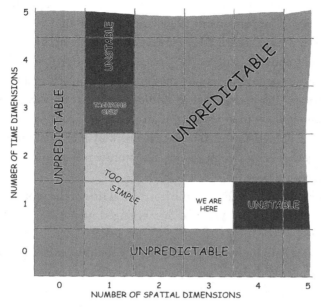

FIGURE 36 Changing the number of spatial and temporal dimensions in the Universe has a significant impact on the basic physics present, and hence the possibility of life. (Based on Tegmark (1997).)

unstable, both in their gravity and their atoms. Let's take a look at the rest of the plot.

What if we removed a spatial dimension? Life in such a two-dimensional world has been considered, and a number of problems are immediately apparent. Three-dimensional beings like us have the topology of a torus,[28] a basic body shape for many animals, with the hole extending from the mouth to the other end. A two-dimensional being that incorporated such an alimentary canal would be split into two separate pieces. Lines cannot cross in two dimensions without intersecting, which is why traffic on the surface of Earth can be such a nightmare. There are ways around this; we could allow our two-

---

[28] It is often stated that a doughnut is a nice example of a torus. Except for British people, whose doughnuts are a slightly squashed sphere, and injected with jam and cream. Yum.

dimensional being to envelop food and digest it like the creature in the superb 1950s (or not so superb 1980s) movie *The Blob*.

But we have more to worry about than whether our two-dimensional cousins can eat. In General Relativity, it is only with three or more space dimensions that gravitational fields can exist in empty space. This means that in two-dimensional universes gravity cannot hold planets in a stable orbit around a star. And, in fact, forming stars and planets in the first place is unlikely if gravity cannot pull matter together across empty space. Such universes would be simple, sterile places.

So far, we have discussed the horizontal line of the plot on which 'WE ARE HERE' lies. What if we changed the number of *time* dimensions?

What would such a universe be like? Clocks would still tick, and so a life form would experience spacetime events in a one-dimensional sequence. We could arrange for an arrow of time to ensure that these events are recorded in the familiar way, creating a one-dimensional sequence of memories.

The difference comes in the effect of these dimensions on the equations of physics. For example, there is an equation in physics that describes waves, whether on a string (one-dimensional), on the surface of water (two-dimensional), or through the air (three-dimensional); it is called, appropriately enough, the *wave equation*. As well as adding space dimensions to our equations, including the wave equation, we can add time dimensions and see what happens.

In our familiar 3+1 dimensions, we rely on information collected from our surroundings to guide our behaviour. We see that there are no cars immediately approaching and so pull out into traffic. A bird can judge the flight path of an insect to catch its dinner. This is so instinctive that we hardly think of it as *prediction*, but that is what it is. In terms of the dimensions of space and time, we use information gathered over a region of *space* to predict what the world will be like later in *time*.

However, with extra time dimensions, we must distinguish between *local* time on our watch (one dimensional) with the *global* time dimensions (zero or more). Suppose our local time lines up with one of the global time dimensions. Then, in order to predict what will happen five seconds into the future (on our watch), we must gather information over a region of *space* and (global) *time*.

The problem: unless you can measure your surroundings with *perfect* accuracy, your prediction of the future will be useless. The addition (or *subtraction* – it's just as bad in zero time dimensions) of time dimensions means that, even though the world around us would obey a simple equation, it would be practically unpredictable. Our observations of the universe around us would give us no information about how it *will be*. The world might as well be totally chaotic. It is unlikely that an information-processing system could exist at all.[29]

So, if our Universe had been born with a different number of spatial and temporal dimensions than the three plus one we currently inhabit, the prospect for complex life arising would have been severely diminished if not downright impossible. This is yet another way in which our Universe appears to be fine-tuned for life.

However, the resolution of this fine-tuning may be found in a speculative theory of fundamental physics, known as *string theory*. According to this theory, if we could look very closely at particles we would see that they are really tiny strings that vibrate in eleven dimensions. To explain why we observe only three space and one time dimensions, string theorists require the other dimensions to be curled up, or *compactified*.

To clarify, the fine-tuned dimensions we discussed above (and in Figure 36) refer to large or *macroscopic* dimensions. The Universe might fundamentally have more dimensions, as in string theory, or even fewer dimensions – in some quantum theories of gravity, there is no time dimension at all, so time must somehow *emerge* from

---

[29] Do you enjoy partial differential equations? Then why not read Tegmark (1997) for a full account of the mathematics behind these arguments!

something more fundamental. To describe our Universe, such theories must have *some* way of producing classical, macroscopic 3+1-dimensional spacetime. The dimensions of spacetime *on the scale of atoms and planets* must be fine-tuned, for the reasons we have explained.

Nevertheless, we are again left with a nagging question. Suppose string theory is correct: was the process that resulted in 3+1 macroscopic dimensions compelled to choose exactly that combination? Or was there an element of chance to it? Could the universe have been born with a different mix of space and time?

String theory is silent on the matter, since after nearly forty years of scientific and mathematical investigation, it is an incomplete (and notoriously difficult) theory. But it could be that a cosmic roll of the dice decided the dimensionality of space. If the dice are rolled enough times, a 3+1-dimensional universe is bound to turn up somewhere. But we are getting ahead of ourselves – we'll return to string theory, compactification and other universes in Chapter 8.

MATHEMATICAL SPACES

Near the start of this chapter, we discussed the importance of the symmetries hidden in the mathematical laws of physics. Perhaps surprisingly, the quantum world of the electron and photon and the curved and expanding spacetimes of the entire Universe are both described in terms of a unified, underlying mathematical language. This language was discovered when the seventeenth-century French scientist Pierre de Fermat was thinking about light.

When we first hear about light in our school science class, we learn that it travels in straight lines through empty space, and will reflect off mirrors. But think about gazing at the image of a flickering flame in a mirror. Just which path does light travel from the flame to the mirror to your eye?

Fermat's answer was that light follows the path that takes the least time between the flame, the mirror and your eye. Because light travels at a constant speed through air, this is the path of least distance; all of the other possible paths are longer. Fermat's 'principle of least time' shows its real power when it is applied not only to light's reflections, but to all of light's journeys.

When light travels from air into water, its path is bent. *Snell's law* describes this effect, though it was known long before it was explained by its namesake, the seventeenth-century astronomer Willebrord Snellius. This bending is a consequence of light moving more slowly through the water. When we consider all of the possible paths light could have travelled from a source above the water to the eye of an observer below it, the actual path is the *quickest*.

Since Fermat, this notion of the 'path of least time' has become central, not only to the motion of light, but all of physics. When we peer into the mathematical framework of the pillars of modern science, be it electromagnetism, quantum mechanics or Einstein's General Theory of Relativity, we see a more general but similar 'principle of least *action*' at work. In this case, *action* doesn't refer to the constant partying that physicists are so famous for, but to a mathematical quantity that succinctly but powerfully encodes the laws of nature.

At university, a physics student encounters this 'least action principle' in *Lagrangian* and *Hamiltonian* approaches to classical mechanics.[30] The symmetries we discussed previously, which entail the conservation laws of the Universe, are most useful within these mathematical frameworks (Figure 37).

Intriguingly, beneath the very different ideas about the universe found in Newtonian physics, Maxwell's electromagnetism, quantum mechanics and Einstein's relativity, we find a common core. They all

---

[30] We heartily recommend *Classical Mechanics: The Theoretical Minimum* by Susskind and Hrabovsky (2014) for anyone wanting to truly get their hands dirty with the principle of least action. A little maths, and a lot of thought, is required, but you will get a deep understanding of how physics is done.

Physics is often taught in chronological, historical order, rather than presenting the most relevant tools up front. It's a little like starting an apprentice in car mechanics off by getting them to hit rocks together. We do hope science education will modernize!

$$S = \int L\, dt$$

$$\frac{d}{dt}\left(\frac{\partial L}{\partial \dot{q_i}}\right) - \frac{\partial L}{\partial q_i} = 0$$

FIGURE 37 The action equation and the Euler–Lagrange equation. These are truly the central equations of modern physics, appearing in both relativity and quantum mechanics, and our attempts to breech the chasm between them.

boil down to a principle of least action. At the most fundamental level, the physical rules that govern the workings of the Universe appear to be built upon the same mathematical foundation. But why?

Ultimately, we don't know. We are reaching 'why?' questions that physics cannot hope to answer. Could the Universe have been born with a different mathematical heart? And if it could, what would it be like?

It is difficult to speak generally about universes based on alternative mathematical structures, since there are so many possibilities. Indeed, this is why fine-tuning has mostly considered changes in the constants of nature – at least we know the laws.

However, stranger worlds are available. There are alternative universes that are simple enough for us to explore in some detail. Indeed, mathematicians have been frolicking in these virtual universes for many decades.

LIFE ON THE GRID

Our Universe's space and time are *continuous*. So far as our experiments can tell, between any two different points in space or time, we can find another point. Beginning in the 1940s with Stanislaw Ulam and John von Neumann, mathematicians have also investigated 'universes' that are not continuous, but rather are *discrete*. Think of a chessboard: in the game of chess, there simply isn't anywhere between squares a1 and a2. Mathematically, rather than labelling

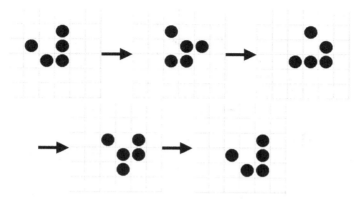

FIGURE 38 A 'glider' in Conway's Game of Life. The cells with black dots
are *on*, and the white cells are *off*. The pattern in the top right, over the
course of four time steps, moves one space down and to the right.
The pattern appears to glide diagonally on the grid.

places and times with decimals, we label them with whole numbers:
1, 2, 3, ...

Having created a discrete universe, we can create some discrete
entities that obey discrete rules. One particularly splendid example
was invented in 1970 by mathematician John Conway, and is known
as the *Game of Life*.[31] (It's not really a game. It's a simulation.)
It works as follows.

The universe of the Game of Life is a rectangular, two-
dimensional grid of *cells*. The cells are very simple: each is either *on*
or *off*. The rules are similarly simple. At a particular time, and for each
cell, count how many of its eight neighbouring cells are *on*. If,

A. the cell is *off*, and 3 neighbours are *on*, or
B. the cell is *on*, and 2 neighbours are *on*, or
C. the cell is *on*, and 3 neighbours are *on*,

then the cell will be *on* at the next time step; otherwise, the cell will
be *off*.

[31] Originally published in Gardner (1970).

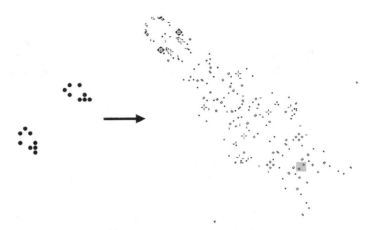

FIGURE 39 Noah's Ark, in Conway's Game of Life. The starting pattern on the left, with just 15 *on* cells, grows after 2,688 generations into the pattern on the right. The grey region on the right shows the location of the starting pattern. This 'puffer' will continue to grow diagonally up and to the left.

We start a universe with a configuration of *off* and *on* cells, and then press 'go'. The universe will play out, cells will live and die, and shapes and patterns will grow and decay. For example, Figure 38 shows a shape known as a *glider*. Over four time steps, the shape moves down and to the right.

Fantastic beasts have been discovered in Conway's universe. There are all manner of blinkers, beacons, breeders, glider guns, pulsars, puffers, spaceships, and more. In 1971, Charles Corderman discovered 'Noah's Ark', which begins in a region of 15 by 15 cells, 16 of which are on (shown in Figure 39), and grows to produce a diagonally repeating pattern (a 'puffer') that every 1,344 generations leaves in its wake '42 blocks, 40 blinkers, 22 beehives, eight loaves, four gliders, two boats, two block on tables, two long boats, one ship, and one beacon'.[32] To see it in action, download a program from *conwaylife. com* to run the Game of Life on your computer.

---

[32] From the 'Noah's ark' entry at *conwaylife.com/wiki/*.

Importantly, and rather astonishingly, within the Game of Life you can build a *universal Turing machine*. That is, a carefully built entity in Conway's universe can simulate *any* computer algorithm.[33] It's hard to know what to call 'life' in the grid, but we can at least say that Conway's universe has a lot of software at its disposal.

Universes like Conway's, where discrete rules play out on a grid, are known as *cellular automata*. Alternatives to Conway's universe consider different rules, more than two states for each cell, or even higher dimensional grids. The number of possibilities is truly vast. The Game of Life's cousins – that is, games that are two-dimensional, have two states (*on/off*), and in which the rules about the next time step depend on the current state of a cell's neighbourhood – number more than 10^{154} different possible rules. Even if the rule only depends on the *number* of *on/off* neighbours, there are still 262,144 rules.

In this book, we have set ourselves the task of investigating other possible universes, and so cellular automata are within our purview. But note that it will not suffice to point out interesting examples of cellular automata, such as Conway's. We have expanded our search area. The question is whether we will find an Eden, with complexity and life in every corner, or a desert, with life huddled in a few tiny oases.

We would love to report the fraction of cellular automata that support the kind of balance between simplicity, complexity and novelty that we see in Conway's universe. Unfortunately, there is no simple way to determine this. Even the more precise question of which permit a universal Turing machine is too difficult to answer *en masse*. However, a few pieces of evidence suggest that interesting cellular automata are rare.

---

[33] This section will not do justice to the mathematical rigour of the work of Turing and his colleagues and successors in computer science and information theory; see *The Information: A History, A Theory, A Flood* by James Gleick (2012) for an introduction to this world.

The first is the development of Conway's Game of Life. The simplicity of the rules didn't come for free. Conway and his students searched for rules that produced an unpredictable universe, and took some of the ideas for rules from real biological systems: the rules implement 'death' by isolation or overcrowding, and life and birth in between. In Richard Guy's (2008) short biography of Conway, he comments:

> His discovery of the Game of Life was effected only after the rejection of many patterns, triangular and hexagonal lattices as well as square ones, and of many other laws of birth and death, including the introduction of two and even three sexes [states]. Acres of squared paper were covered, and he and his admiring entourage of graduate students shuffled poker chips, foreign coins, cowrie shells, Go stones, or whatever came to hand, until there was a viable balance between life and death.

That sounds a lot like fine-tuning – within a vast space of possibilities, only a very small fraction show an interesting balance between boredom and bedlam.

The second hint is more formal. Mathematicians have tried to categorize cellular automata, noting four types of universes that result from a certain set of rules. (Stephen Wolfram, who advocates that complex, physics-like laws can be generated by cellular automata, proposed these classifications in the mid-1980s.) While not totally rigorous – there are cases that are difficult to classify – they are useful.

Class 1: boring. The universe tends to stable uniformity. Every cell is either on or off, and soon nothing ever changes.

Class 2: repeating. The universe tends quickly to simple stable or periodic patterns. Cells stay on, stay off, or blink.

Class 3: chaos. There are no stable structures, as any pattern quickly dissolves back into the noise.

Class 4: interesting. Complex patterns abound, with stable structures surviving, moving and interacting for long periods of time.

A good example of a Class 4-type phenomenon is Noah's Ark: a simple starting pattern explodes into a complex assortment of shapes. More than a thousand time steps pass before it settles into a complex, repeating pattern.

Class 4 has attracted most interest from researchers. To create a universal Turing machine in a cellular universe, we need to be able to encode information, protect it, and transport it from one place to another. Classes 1 and 2 will not do, as they are too uniform to encode information, and too stable to transport it. Class 3 is no good either, as the chaos will tend to erase any information we try to encode. Wolfram, in light of this, conjectured that only Class 4 universes could permit universal computation.

Among cellular automata, Class 4 universes are rare. The simplest cellular automata are one-dimensional: just a line of cells, on or off, with the next time step depending on only two neighbours. There are 256 such rules, and only 6 are Class 4. Despite the simplicity of one-dimensional cellular automata, Matthew Cook proved in the 1990s that one of these universes (the famous *Rule 110*, published by Cook in 2004) could indeed contain a universal Turing machine.

Researchers have also noted that a small adjustment to a Class 4 universe often results in either a simple (Class 1 or 2) or chaotic (Class 3) universe. Interesting universes that can support information-processing entities exist in the narrow boundaries between the other classes. As a result, the fraction of Class 4 universes is expected to decrease sharply as spaces with more complex rules (e.g. two and higher dimensions; more possible states per cell) are explored.

These balance points resemble *phase transitions* in physics: for example, when ice is heated, its structure changes abruptly as the temperature reaches melting point. Considering the temperature and pressure of water, there are sharp dividing lines between the phases of water (vapour, liquid, solid). Similarly, in the set of possible cellular automata rules, there are sharp dividing lines between 'too simple' and 'too noisy'. The complex behaviour that mimics life is believed to inhabit the boundaries, the phase transitions in rule space, the *'edge of chaos'*.

In 2014, Sandro Reia and Osame Kinouchi in Brazil investigated some of the 262,144 rules that are close cousins to Conway's Game of Life. They suggest that the interesting rules are found near the boundary that separates universes in which the *off* regions grow from those in which they shrink. Conway's *Life* is very close to this boundary. They conclude, 'complex automata are rare.... [Conway's] LIFE's rule, with its rare property of being a Universal Turing Machine ... is fine-tuned to place the [cellular automaton] at the border of the phase transition.'

Even in the utterly foreign realm of cellular automata, with discrete *'on-off'* cells on a grid instead of fields and particles in space and time, we hear a familiar story. Most of 'rule space' cannot hold, preserve and move information. Patterns are frozen, wiped blank, or overwritten by meaningless noise. But on the edge, if the right balance is achieved, patterns can persevere and propagate, entities can survive and interact, and *any* computation is possible. Interesting, complex, organized, information-processing universes require fine-tuned rules.

ULTIMATE CHAOS REIGNS SUPREME

We have played with the laws and the fabric of the Universe, and the result is generally monotony or mess. But there is one more scenario to consider: a universe with no mathematical rules at all. While this seems extremely bizarre, the fact that we are able to understand the Universe in any sort of ordered sense is, in itself, rather amazing. Einstein stated, 'the most incomprehensible thing about the world is that it is comprehensible.'[34]

Life relies on the fundamentally mathematical nature of our Universe's laws. The countless interlocking mechanisms working in every cell in our bodies depend on the stable, predictable properties of our constituents and the rules by which they interact.

[34] Quoted in Vallentin (1954, p. 24).

The mathematical laws of nature can be thought of as an extra-ordinarily efficient *encoding* of the data of our experiments and observations of the Universe. For example, when electrons jump between permitted orbits in a hydrogen atom, they emit light of a certain frequency. Instead of memorizing the hundreds of observed wavelengths in the NIST Atomic Spectra Database, we can use the *Rydberg formula*. Beneath pages and pages of observation and experiment, we find a simple pattern.

This is not always possible. Mathematicians define the *Kolmogorov–Chaitin complexity* of a string of characters to be the length of the shortest set of instructions that produces it. For example, the sequence,

0101010101010101010101010101010101010101010101

can be succinctly described as '01, 22 times'. By contrast, there is no simple formula for Shakespeare's *Hamlet*, or the following throws of a die,

2136362464625333655165126244523663316365165656.

This provides a powerful, rigorous approach to analysing a famously tricky idea: a string is *random* if there are no instructions shorter than the string itself. In other words, the string is *incompressible*. It is a mathematical theorem of Kolmogorov–Chaitin complexity that most strings are random.[35]

It is thus all the more remarkable that the strings of numbers that represent the results of our experiments and observations of the Universe can be summarized in just a few short equations. The Large Hadron Collider, for example, collects around 50 terabytes of data every day about how particles interact. And yet, the Standard Model of particle physics can be handwritten – in the appropriate mathematical language – on a single piece of paper. Our world can be encoded or

---

[35] '[T]he overwhelming majority of strings have almost no computable regularities'. We have called such a string "random". There is no shorter description of such a string than the literal description: it is incompressible.' Li and Vitany (2008, p. 4).

summarized to an extraordinary degree. The Universe is comprehensible because it is compressible.

What if it wasn't? What if physical phenomena could not be condensed into simple patterns and laws? Such a universe would be truly random, as no event could possibly lead you to expect any other event. Particles, waves, fields, and polar bears might pop in and out of existence with no rhyme or reason. It would be like watching static on an old TV, or reading the collected works of a million monkeys at a million typewriters, hoping for Hamlet.

It is hard to imagine how life could get started in such frantic randomness, and even harder to imagine life thriving, reproducing and evolving. Eventually, within this heaving mess, some apparent order or structure would arise and appear to follow mathematical laws. But this temporary illusion would soon melt back into the maelstrom. Our Universe's laws reflect the order and stability that allow life to exist. Since theoretical physics *starts* with the postulation of such laws, science cannot tell us why the Universe is orderly at all.

And it is with this vision of anarchy that we reach the end of our journey. In considering a universe with no laws, no structure and no predictability, we have arrived at the ultimate chaos.

Throughout our journey, we have encountered a common message. Be it small changes to what we know, or larger changes that shake the foundations of space and time, it seems that the Universe could have been so different, so very dead and sterile.

We find ourselves questioning our existence in a Universe with a nice set of physical laws, with the right masses and forces, the right kind of beginning, played out against a convivial canvas of three dimensions of space and one of time. With so many potential ways the Universe could have been, we cannot ignore the apparent specialness of our existence.

So, what are we to make of this?

7　A Dozen (or So) Reactions to Fine-Tuning

We have presented the science of fine-tuning, as described in previous chapters, to all kinds of audiences – young and old, scientist and civilian, and even the occasional philosopher. It has an uncanny ability to raise questions.

For example, a forty-minute talk to our colleagues at the Sydney Institute for Astronomy – professional astronomers whose job is to understand the Universe – was followed by an hour and a half of questions. For those who have never attended an astronomy seminar: this never happens. In a typical seminar, astronomers become anxious to leave after three or four questions, and for good reason. It's lunchtime.

The fine-tuning of the Universe for life is unique in our experience for the *strength* of the opinions expressed. In both popular and professional settings, disagreements often become noisy arguments. Even those who don't think fine-tuning means anything simply *must* enthusiastically explain to everyone, in great detail, exactly why it doesn't mean anything.

Fine-tuning also evokes a terrific *range* of responses. Most of the audience agrees that this is an unusual kind of fact, and one that mustn't be allowed to ramble free and unexplained. It must be put firmly in its place, either in the service of some conclusion or quarantined away from all such conclusions.

The problem is that different members of the audience can't agree on its proper place. Our neat question-and-answer session following a talk dissolves into a town hall meeting. Questions to the speaker are intercepted by the front row, defended on the left, heckled by the back row, and so on. We just light the fuse and enjoy the fireworks.

Given that fine-tuning is such an odd kind of fact, it will help if we discuss the most common responses we've collected from audiences over the years.

Reaction (a): It's Just a Coincidence

The comment: You'll make yourself mad trying to explain every little detail about the Universe. Some things just are the way they are. If anything needs no further explanation, it's the laws of nature.

The short answer: 'Any coincidence', said Miss Marple to herself, 'is always worth noticing. You can throw it away later, if it is only a coincidence'.[1] Fine-tuning looks an awful lot like a clue to something deeper than the laws of nature *as we know them*.

The long answer: We do not believe that the laws of nature as we know them are the Ultimate Laws of Nature™. In particular, those constants are a liability. For all the mystery and intrigue attached to the (deep, booming voice now) *fundamental constants of nature*, they are free parameters in our equations. They are just numbers that we can't calculate, that we have to measure. They are data in disguise, meter readings that have gate-crashed our mathematical laws of nature. How uncouth!

Any theory with free parameters is liable to be replaced by a theory with fewer free parameters. What will that theory be? Any interesting fact about these free parameters could be a clue to what comes next. Fine-tuning is just such a fact.

Reaction (b): We've Only Observed One Universe

The comment: Let's review what we know. How many universes have we observed? One. Of the universes that we have observed, how many contain life? One. So there's your probability: one out of one universes contains life. Where's the fine-tuning?

The short answer: Physics is about more than observations; it is also about patterns, theories and laws. It's about exploring the 'what

---

[1] From *Nemesis* by Agatha Christie (1971).

ifs'. The universes of fine-tuning are theoretical universes that explore the possibilities left open by the laws of nature as we know them. So we're comparing *theoretical* universes; the number of *observed* life-permitting universes isn't the whole story.

The long answer: Physicists have discovered that a small number of simple mathematical rules can account for mountains of data that we have collected about how our Universe and its contents behave.

These laws, however, describe more than the way the Universe actually is. Newton's law of gravitation, for example, describes the force of gravity between *any* two masses separated by *any* distance. *This* feature of the laws of nature makes them predictive – they not only describe what we have already observed; they place their bets on what we observe next. The laws we employ are the ones that keep winning their bets.

Part of the job of the theoretical physicist is to explore the possibilities contained within the laws of nature to see what they tell us about the Universe, and to see if any of these scenarios are testable. For example, Newton's law allows for the possibility of highly elliptical orbits. If anything in the Solar System followed such an orbit, it would be invisibly distant for most of its journey, appearing periodically to sweep rapidly past the Sun. In 1705, Edmond Halley used Newton's laws to predict that the comet that bears his name, last seen in 1682, would return in 1758. He was right, though didn't live to see his prediction vindicated.

This exploration of possible scenarios and possible universes includes the constants of nature. To measure these constants, we calculate what effect their value has on what we observe. For example, we can calculate how the path of an electron through a magnetic field is affected by its charge and mass, and using this calculation we can work backwards from our observations of electrons to infer their charge and mass.

Probabilities, as they are used in science, are calculated relative to some set of possibilities; think of the high-school definition of probability as 'favourable over possible'. We'll have a lot more to say about probability in Reaction (o); here we need only note that scientists test their ideas by noting which possibilities are rendered probable or improbable by the combination of data and theory. A theory cannot claim to have explained the data by noting that, since we've observed the data, its probability is one.

Fine-tuning is a feature of the possible universes of theoretical physics. We want to know why our Universe is the way it is, and we can get clues by exploring how it could have been, using the laws of nature as our guide.

Reaction (c): Low-Probability Events Happen All the Time

The comment: Forget about life in general; let's consider the probability of *you*. Think of all the coincidences involved in your parents being in the same place and meeting and hitting it off and getting together. Think of the miniscule probability of a particular sperm outracing a billion others to find the egg. Multiply similarly tiny probabilities for all your ancestors stretching back in time, and you get an extraordinarily small probability. Yet, here you are. You're just going to have to get used to it.

The short answer: Small probabilities sometimes mean that something unlikely has happened. Enough said. But sometimes they mean that we've made an incorrect assumption. Given that almost no one believes that the laws of nature *as we know them* are the ultimate laws of the Universe, the low probability of a life-permitting universe could be a clue to a better explanation, a deeper theory.

The long answer: Sure, the lawyer says, the DNA evidence makes it extraordinary unlikely that my client is innocent. But, your honour, unlikely events happen all the time! Eggs and sperm and such! The defence rests.

Something must have gone wrong with this response to fine-tuning, since the same reply could be made to any appeal to low probabilities. We'd never be able to reason probabilistically at all.

Think about some seemingly improbable events: a poker player deals himself another royal flush, a large blip appears on our detector, a safe with a trillion possible combinations is opened. In these cases, a small probability is generated not just by the event but also by our *assumptions*. We've assumed that the dealer is fair, that the instrument reading is just noise, or that the burglar guessed the combination to the safe.

What separates these from the 'just unlikely' cases is the availability (or even just a glimpse) of a *better explanation*: a trick shuffle, a signal, or an inside job. This is precisely what we don't have in the case of 'egg + sperm = you'.

So before we dismiss a low-probability event as just a fluke, we should consider alternative explanations. What are the deeper, better explanations that fine-tuning suggests? Stay tuned!

Reaction (d): Fine-Tuning Has Been Disproved By (Insert Name Here)

The comment: A number of scientists have looked closely at these fine-tuning claims and concluded that they are at least dubious, if not false. Fine-tuning has been debunked.

The short answer: No. It hasn't.

The long answer: Luke has published a review of the scientific literature on fine-tuning, carefully summarizing the conclusions of over 200 published papers in the field. These papers have built on the original work of key physicists, Carter, Silk, Carr, Rees, Davies, Barrow and Tipler, who pioneered the field. Their calculations have been refined using cutting-edge models and methods. Sometimes, new options for life have opened up, and sometimes life has turned out to be more fine-tuned than previously thought. On balance, the fine-tuning of the Universe for life has stood up well under the scrutiny of physicists.

A good example of the way this field has progressed is the work of Ulf-G. Meißner and collaborators. His team of experts in the theory of nuclear forces turned their attention to the Hoyle state of the carbon nucleus. As we saw in Chapter 4, this state plays a crucial role in the production of carbon in stars, and hence the manufacture of the backbone of DNA and a host of other biological building blocks.

In 2011, they used supercomputer simulations to show that nuclear theory can correctly predict all the measured properties of the Hoyle state. In 2013, they used their state-of-the-art model to investigate the 'Viability of carbon-based life as a function of the light quark mass'. They calculated that changes in the masses of the quarks or the strength of electromagnetism of more than a small percentage significantly affect the ability of the universe to produce both the carbon *and* oxygen required by organic life.

In many examples such as this one, our best theoretical physics has informed our knowledge of the sensitivity of the conditions for life on the fundamental constants of nature.

Only a handful of peer-reviewed papers have challenged the fine-tuning cases we've discussed in this book,[2] and none defend the contention that most values of the constants and initial conditions of nature will permit the existence of life. While there is more work to be done, and in particular much more to be learned about the formation of life, the progress of theoretical – and particularly computational – physics has tended to consolidate our understanding of fine-tuning.

This reaction might stem from the belief that fine-tuning is the invention of a bunch of religious believers who hijacked physics to their own ends. This is not the case: the field began in physics journals and remains with physicists such as Barrow, Carr, Carter, Davies, Deutsch, Ellis, Greene, Guth, Harrison, Hawking, Linde, Page,

---

[2] Steven Weinberg argued on plausible grounds that the Hoyle state would have life-permitting properties for a large range of parameters; this has since been shown to be incorrect by the calculations of Epelbaum et al. (2011, 2013) and Meißner (2015). In Chapter 3, we discussed Harnik, Kribs and Perez (2006), who argued that life was possible in a universe with no weak force. This required 'judicious parameter adjustment' (their words) of the other parameters, and might not create sufficient oxygen. Aguirre (2001) argued that a cold Big Bang – one in which there were roughly the same number of photons as particles of ordinary matter – could be life permitting. As discussed in Chapter 6, the relevant parameter (known as η_γ) depends on the physics that makes matter outnumber antimatter in our Universe, physics that is largely unknown. All of these cases are discussed in Luke's review paper.

Penrose, Polkinghorne, Rees, Sandage, Smolin, Susskind, Tegmark, Tipler, Vilenkin, Weinberg, Wheeler and Wilczek.[3]

Reaction (e): Evolution Will Find a Way

The comment: Life, we have discovered, is extraordinarily resilient. It has adapted to extreme environments, from Antarctic ice to undersea volcanic vents. The ability of life to survive in extreme conditions of heat, cold, acidity, radiation, pressure, dryness, and nutritional scarcity is testament to its hardiness. Evolution will adapt life to whatever conditions it finds.

The short answer: Even the most extreme conditions on Earth are a paradise compared to the horrors that await in most of parameter space. Forget the thermostat; we'll dissolve your atoms and crush your particles in black holes or big crunches. Further, evolution assumes a minimum of biochemistry: until you have a life form that can create copies of itself according to some internal plan, Darwinian evolution can't even get started.

The long answer: Organisms on Earth can survive in an impressive range of conditions. The toughest life forms, known as *extremophiles*, have adapted to survive and flourish in environments that would wipe out most other life. For example, the imposingly named *Methanopyrus kandleri strain 116* is a single-celled microorganism that grows at temperatures of 122 °C and pressures 200 times higher than those we experience. Meanwhile, there is a small insect – a type of midge – living in a high-altitude glacier in the Himalayas that is active at temperatures so cold (−16 °C) that even insects from the Antarctic would put on a coat and go home.

Have we underestimated life? In the words of the world famous mathematician, Dr Ian Malcolm:

> If there is one thing the history of evolution has taught us, it's that life will not be contained. Life breaks free, it expands to

---

[3] References in Barnes (2012).

new territories and crashes through barriers; painfully, maybe even dangerously ... I'm simply saying that life, uh, finds a way.

Admittedly, Dr Ian Malcolm is fictional – read the quote again in your best 'Jeff Goldblum smugly predicting the fate of Jurassic Park' voice – but the point is well made.[4]

But there are two problems with this reaction. The conditions faced by life on Earth aren't remotely representative of conditions in the parameter space of other universes. They aren't even representative of conditions in our galaxy. Take temperature, for example. Life can survive and remain active between –20 °C and 122 °C. Most of interstellar space is hot,[5] with temperatures between 6,000 °C and 10,000 °C, except for dense molecular clouds at –260 °C. Moreover, interstellar space is a trillion times less dense than even the most tenuous regions of our atmosphere in which microbes have been found. This makes Earth look balmy.

At no point did we conclude that an alternative universe was life-prohibiting because of its pressure, saltiness or acidity, or because its temperature would raise a sweat in even the roughest, toughest microbe. The extremes of parameter space are not just hot and cold. Rather, they are disintegrating atoms, the cessation of all chemical reactions, the crush of a black hole, and the eternal loneliness of life in a universe where particles collide every trillion years or so.

The second problem is that extremophiles are *adaptions*. They show the conditions in which life can *live*, not in which life can *form*. Given the adaptions that life has managed in the last few zillion generations, it is likely that the conditions under which life formed

---

[4] You may find yourself reading the next few paragraphs in a Jeff Goldblum voice. This is, uh, perfectly normal.

[5] That might surprise you, as we usually think of space as cold. The temperature of the matter of interstellar space is set by a balance between heating and cooling processes. In such diffuse environments, the processes that cool the gas are very inefficient below about 5,000 °C. The combined effect of stellar radiation, cosmic rays, and supernovae keeps the gas very hot. However, because it is so thin and diffuse, it would not *feel* hot if you – unwisely – stuck your head out of your spaceship's window.

on Earth are much more temperate than the extremes to which it has since spread.

This highlights the problem. These extremophiles show us what evolution can do. But that was never the issue. Rather, the question is what a universe needs to do in order for life to form and evolve *at all*. While a single cell is a simple thing in biology, it is a miracle of chemistry. And chemistry requires a lot from physics. It takes a well-adjusted universe to produce and assemble all these chemicals on a planet at a safe distance from a stable source of energetic photons that can power photosynthesis, and so on.

The powers of Darwinian evolution do not come for free. A biochemical machine, able to draw energy and nutrients from its environment for the purpose of making a complete working copy of itself, including an almost-but-not-exact copy of its genetic instructions, in an environment in which there is competition for resources, is where evolution *starts*. The apparently rare ability of our Universe to bring forth such creatures, capable of Darwinian evolution, is precisely what we are trying to explain.

Reaction (f): How Can the Universe Be 'Fine-Tuned' When It Is Mostly Inhospitable To Life?

The comment: Fine-tuned? You must be kidding: 99.99999... per cent of the Universe is radiation-filled vacuum. Most of the matter is inhospitable: suffocatingly diffuse gas that is either unimaginably cold (–260 °C) or roastingly hot (1,000,000 °C), thermonuclear stars, matter-crushing black holes, to say nothing of the occasional supernovae or gamma-ray burst. The Universe is mostly inhospitable, and the parts that are hospitable are very inefficient at creating life. This Universe sure doesn't look fine-tuned for life.

The short answer: This reaction compares conditions on Earth with the conditions elsewhere in this Universe, and thus completely misses the point. We want to know why this Universe has the fundamental properties that it has. We thus compare our

Universe with other possible universes, not just different locations in this Universe. Further, the size and relative emptiness of the Universe is not irrelevant to life. Smaller, denser universes tend not to last very long.

The long answer: 'Life-permitting' does not mean 'crammed with living beings from end to end and from start to finish'.[6] It does not mean that every time and place in this Universe could support life. It doesn't mean that you can set up your deckchair wherever you like and expect a cocktail.

The observation that the Universe is fine-tuned for life, then, does not solve the *Fermi Paradox*, which poses the question *where is everybody?* If Earth has evolved intelligent life forms that may soon be able to travel beyond the Solar System and colonize the galaxy, then why do we see no evidence of other alien civilizations? Are we the first? Are we alone in the Galaxy? Do intelligent civilizations self-destruct before they leave their home planet? Fine-tuning only tells us that our Universe has some of the necessary physical conditions for life. It doesn't tell us whether life will actually form in every nook and cranny of the Universe.[7]

Remember why we started poking around in the laws of nature in the first place: our equations look unfinished, and contain constants that we cannot calculate. Why is our Universe like this? We looked for clues by comparing our Universe with its siblings, distant cousins and more. We found that very few family members were able to evolve and sustain the kind of intelligent life forms who would ask such questions in the first place.

Parts of this Universe don't support life, but that is not the big story. Indeed, if you can understand why life cannot exist in the near-vacuum of interstellar space, then you can understand why life needs a fine-tuned universe. If the cosmological constant was slightly larger or the cosmic inhomogeneity (Q) smaller, *only* the vacuum of

---

[6] This quote is from John Leslie's *Universes* (1989), p. 159. Required reading.

[7] Of course, this is not to say that the question of life elsewhere in this Universe is not a fascinating and important one; see *The Eerie Silence* by Paul Davies (2010).

interstellar space would exist, without any stars or planets at all! Inhospitable matter is easy; that this Universe has hospitable conditions *anywhere* is an extremely rare feature.

Further, 'fine-tuned' does not mean that our Universe packs the maximum amount of life into a given region, or makes all the life that can be made out of a given set of particles. Perhaps there are other parts of parameter space that make life more efficiently. For example, it is plausible that a universe with zero cosmological constant (as opposed to our small cosmological constant) puts more of its ordinary particles into galaxies than ours does, and so makes life slightly more efficiently than ours.

So it might not be the case that our Universe is optimally packed, cheek by jowl, with life. But, again: this is missing the headline! Consider the vast landscape of possible universes. It is dominated by wastelands of dead, boring, simple, lifeless, unobservable universes. Rare, narrow, clustered mountains of life rise sharply from the mire. Perhaps, if we get out the magnifying glass, we will find that we are not exactly at the apex of our spire. This does not excuse us from explaining our highly unusual position. We are still in possession of a highly unusual, seemingly-explainable-but-unexplained fact.

Indeed, why should we care that other parts of the Universe aren't so accommodating? Suppose that you are working behind the reception desk of a luxurious mountaintop resort. Bob, a wealthy client, is checking in.

YOU: You'll be in room 401, sir. The Penthouse. It has wonderful views of the entire mountain range, and on a clear day you can see all the way to the ocean.

BOB: Oh, that's no good. I don't want to be able to see the ocean.

YOU: Why not?

BOB: Because I can't swim.

Now, being well trained and polite, you book Bob into another room. But on a more sarcastic day, you might be tempted to ask why it would be a problem that, from his room, Bob can see somewhere – miles and miles away – that would be rather uncomfortable for him.

How does that ruin the room? He'd have to invest a considerable amount of time, resources and effort just to get near that dreaded ocean. Even if Bob has a morbid fear of drowning, that's no reason to give the room a bad review on *ratethathotel.com*.

Similarly, why would anyone live in Australia? It's mostly desert![8] True, but some of the non-desert parts are rather lovely.

You can't survive in space. What is your business? You're on Earth. You're positively stuck down here with the ample supplies of oxygen and water. It would take a considerable amount of time, money and technology to experience the discomforts of the vacuum.

In fact, we can glimpse reasons why a relatively empty universe, even if not totally *necessary*, isn't too *surprising* to life. It has to do with the density needed by life, and gravity's crush.

Consider a ball of matter. The collective attraction of gravity will attempt to crush it. Without another force to oppose gravity, how long will it survive? It turns out that this only depends on the density of the ball – the more dense, the more quickly gravity crushes. Even if the ball is huge, at a fixed density, that extra size just means more mass for gravity to use.

So, perhaps you'd like the entire universe to contain a breathable density of air. The problem: the collapse time for a system with such a density is about 24 hours. Beautiful one day, crushed into oblivion the next.

Now, you could escape this collapse by making the system expand, as our Universe expands. The problem is that *one day* is now the typical amount of time it takes for the density of the universe to change significantly. So around this time tomorrow, that breathable air will be rather too rarefied.

---

[8] 'A ridiculous place,' complains the thoroughly Irish Dylan Moran, 'located three quarters of a mile from the surface of the Sun, people audibly crackling as they walk past you on the street. ... It's not supposed to be inhabited, and when they're not frying themselves outside, they all fling themselves into the sea, which is inhabited almost exclusively by things designed to kill you; sharks, jellyfish, swimming knives, they're all in there.' From *Like, Totally ... Dylan Moran Live* (DVD, 2006); used with permission.

Incidentally, the air in this room escapes this fate by being supported by the Earth. The Earth doesn't collapse because, thanks to the rigidity of atoms, it is supported by its internal pressure. Objects supported in this way against gravity can't be much larger than Earth, and must be surrounded by empty space.

Lengthy Aside: A Rotating Universe? What if, taking a hint from our Solar System, we stabilized the whole universe by making it rotate? While this might be a viable option if our Universe were governed by Newtonian gravity, in Einstein's gravity things get weird. Very weird. Weird even for Einstein.

In 1949, Einstein celebrated his 70th birthday at a conference in his honour. In attendance was one of his closest friends, Kurt Gödel, the famous, and famously eccentric, Austrian logician. Gödel is known for his revolutionary work on the logic of mathematics, but before 1949 had not contributed to the world of physics. He brought Einstein a most unusual gift: a solution to the laws of Einstein's gravity that describes a rotating universe. This universe contains only matter and a cosmological constant, and apart from its rotation does not change with time.

This seems ideal: just fill it with air and we could breathe easy anywhere in the universe. However, Gödel's universe has a strange feature: *closed timelike curves*.

Remember that in a relativistic universe you cannot travel faster than light. A hypothetical flight-path through space and time that has you leaving at 5 p.m. to pick up the laundry from Alpha Centauri and being home for dinner is not one that any spaceship could actually follow. Paths that you could travel along – given an adequate rocket pack – are known as *timelike*.

Further, a *closed* curve is a loop; it returns to where it started. So a closed *spacetime* curve returns not just to *where* it started, but also *when* it started. It is a trip that comes back to the same place *and time*.

So, putting together closed and timelike … *time travel!* In a rotating universe, you can visit the same time and place more than once. You could play a game of poker with yourself.

To the science fiction writer, this all sounds great, but beware the infamous *grandfather paradox*. What if I depart the present in my spaceship, travel backwards in time, and crash into my own grandfather when he was just a lad! Then I am never born, and never travel back in time, and never kill my grandfather, and so then I'm born, and then I travel back, and crash . . .

Not all time-traveller-permitting universes are paradoxical, so long as we take care to construct a consistent history. If my spaceship narrowly misses my grandfather, then the paradox is avoided. In such a universe, time travel would not *change* the past: it is not that my grandfather has his original life, and then I am born, and then I go back, and then my grandfather has his 'narrowly missed by a spaceship' life, which overwrites the original. There is only one past, one timeline, in which my grandfather has a narrow miss with a spaceship piloted by a rather familiar-looking young fellow.

But this kind of consistent history requires a lot of arrangement. One cannot simply start the universe off and let the laws of nature play out. To avoid paradoxes, a very tight leash is needed.

It's not clear how agents like us could live in such a universe. Our actions would be constrained by rules and regulations that, unlike the laws of nature, would not be local and discoverable but global and baffling. One would need to know everything about the entire history of the universe to know why any part of it behaves as it does. For example, perhaps I could brush my teeth yesterday, but find myself strangely unable to today. Unbeknownst to me, this is because the cosmos must conspire to prevent the photons that would have bounced off the toothpaste and out the window from travelling around the universe, back into the past and distracting my youthful grandfather while he is driving, causing a fatal car crash.

We included this aside for two reasons. Firstly, because Gödel's rotating universe is awesome. Secondly, this universe-building business is harder than it looks. Some very innocent ideas – such as rotation – can have very interesting consequences. We need to go back to the equations; which is what fine-tuning does.

End of Aside Filling the universe with breathable air has a number of other potential problems. A particle of light will travel about a hundred kilometres in air before it is scattered. Thus, in a universe filled with breathable air, we wouldn't be able to see the Moon, let alone anything beyond. We would live in a fog. Energy from the Sun would not reach the Earth directly, but would have to heat the intervening air.

Further, space travel would be very slow and inefficient. Instead of being able to sail through near-empty interstellar space, spacecraft would have to plough slowly through all that air, burning fuel all the way. You might be able to live with the impossibility of space travel, but the Earth cannot: air resistance would send it spiralling into the Sun within a few months.

We could go on, but it suffices to say that this Universe is not a waste of space. The vacuum, believe it or not, plays its part in making our Universe life-permitting and discoverable.

Reaction (g): This Universe Is Just As Unlikely As Any Other Universe

The comment: Sure, this Universe is unlikely. But any universe is unlikely. We had to be in one of the many possible universes, and so our Universe will always look improbable to some degree. So there is no reason to draw any conclusion from the improbability of our Universe.

The short answer: What matters is not the probability of this Universe compared to the probability of some other universe. Rather, we should be considering the probability of this Universe *given* different theories about how this Universe came to be this way. We are using what we know about this Universe to test ideas about its laws and origin.

The long answer: This reaction presupposes a general principle: where two outcomes are equally probable given some hypothesis, the observation of one of these outcomes rather than the other is no reason to reject that hypothesis. So, for example, if Bob and Jane are equally likely to win the lottery – assuming that they both played fairly – then the fact that Jane wins is no reason to suspect her of cheating.

While this particular conclusion is correct, the general principle is false. To show this, we need only produce a counterexample. It wouldn't be a *general* principle if it failed every now and then.

Here is the counterexample. Suppose that Bob and Jane are playing poker. On the last five hands that Bob dealt, Jane has received middling hands; a pair, King high, a pair, another pair, and Jack high. Bob, meanwhile, has dealt himself royal flush, royal flush, royal flush, royal flush, and royal flush.

Suspicious, thinks Jane. The probability of five royal flushes in a row, given that Bob is dealing fairly, is about one in a hundred billion billion billion. But, seeing her raised eyebrow, Bob protests: his set of five hands is just as improbable as Jane's set.[9] Any set of hands is unlikely. So there's no reason to think that Bob is cheating.

Clearly, something has gone wrong here, and this is enough to conclude that the general principle is wrong. In fact, see exactly where we've gone wrong. The problem is that we have calculated all our probabilities on the assumption that *Bob is dealing fairly*. But that's precisely what we're doubting! What we should be doing is comparing the probability given that Bob is dealing fairly with the probability given that Bob is cheating.

Given that a royal flush cannot be beaten, it is much more likely to be dealt by a cheater than by a fair player. On the other hand, Jane's set of hands provides no particular advantage, and so is about as likely to be dealt by a fair dealer as by a cheater. It is the difference between these probabilities that implies that Bob is probably cheating.[10]

Applying this to fine-tuning, the statement 'this Universe is as improbable as any other universe' is only true if we assume that the

---

[9] Be sure to calculate the probability of each exact hand (e.g. K♥, Q♠, 10♠, 5♦, 4♦), not just 'some King high hand'.

[10] To forestall a common objection, we remind the reader that this is a *counterexample*, not an *analogy*. We are not saying that the Universe is like a poker game. Disanalogies between universes and poker games – 'ah, yes, but card games have rules' – do not overturn the counterexample.

properties of our Universe are random, or are simply brute, unexplained facts. These are precisely the ideas that we are examining. We should consider alternative explanations of the properties of our Universe.

Reaction (h): How Do We Know What Would Happen In Other Universes? Go Do the Experiment!

The comment: Scientific theorizing is fine, but the real power of science comes from *testing* its theories. We make observations and do experiments to collect data. We then compare these data with our theories to see which can explain the data and which are ruled out. All this talk of other universes is impossible to test with actual data, and so fine-tuning is just idle speculation, not science.

The short answer: Comparing our theories to the data of our Universe is one part of physics, and an indispensable part. But to do this, we need to come up with theories in the first place! We would particularly like to find simple, elegant theories with few unexplained assumptions. Finding a theory with fewer loose ends – that still explains the data – is often how physics progresses. Thus, we ask: is there anything noteworthy or unique about the unexplained parts of the laws of nature as we know them that could hint at something deeper? Fine-tuning looks like just such a clue.

The long answer: Exploring our scientific theories is one of the most important tasks of the theoretical physicist. For example, Einstein's published his General Theory of Relativity in 1915, and we are still trying to work out all of its implications.

Once we have the equations of a theory, we can try to solve them. Solutions to the equations of physics represent possible universes. We can use data to try to find the actual Universe amongst the possibilities, but there is more to theoretical physics than this. We believe that successful theories tell us something about our Universe. What does General Relativity tell us about spacetime, about the inside of black holes, and about the very early Universe?

In the past, physics has progressed by noticing something unusual about the laws of nature as we know them. Sometimes, they simply don't match new observations. But sometimes, in order to match the data, we have been forced to make a suspicious assumption.

Gravity provides a great example. When Apollo astronaut David Scott dropped a hammer and a feather on the Moon, where there is no air resistance, they hit the ground at the same time. Why?

In Newtonian gravity, there are two different concepts that we call *mass*. There is *inertial* mass, which measures how hard something is to push. Also, there is *gravitational* mass, which measures how hard something is to lift. Conceptually, inertial mass has no necessary connection with gravitational mass. To measure inertial mass, we could use any force (a magnet, for example) and record how the object responds.

And yet, to explain cases like the hammer and the feather, Newtonian gravity must assume – for no deeper reason – that for every object in the Universe, inertial mass and gravitational mass are exactly equal. This is a bare postulate. Not impossible, not in conflict with observations, but suspicious.

Einstein, thinking purely theoretically, showed the way forward. Newton thought of gravity as a force, which made objects travel along curved paths. What if, thought Einstein, gravity changed geometry itself? What if gravity made objects travel along straight lines in a curved space? Then, because the path of the object would depend only on the local properties of space and time, objects of different mass would fall at the same rate. Einstein explains the hammer and the feather without any additional, suspicious postulates.

The moral: when a theory requires very precise but unexplained postulates to explain the data, it might be a clue. Perhaps we should be looking for a deeper theory in which the postulate is explained, or arises naturally.

The existence of life in our Universe is just such a suspicious fact. It requires a range of precise coincidences between the unexplained parts of the laws of nature: the constants and initial conditions. Can we find a plausible, deeper theory that explains why our Universe is life-permitting?

Reaction (i): Fine-Tuners Turn Only One Dial at a Time

The comment: All these fine-tuning cases involve turning one dial at a time, keeping all the others fixed at their value in our Universe. But maybe if we could look behind the curtains, we'd find the Wizard of Oz moving the dials together. If you let more than one dial vary at a time, it turns out that there is a range of life-permitting universes. So the Universe is not fine-tuned for life.

The short answer: This field started when physicists noted coincidences between the values of *a number of different constants* and the requirements for life. Life requires a number of different constants to be related to each other in unusual and precise ways.

The long answer: This is a surprisingly persistent myth, and one with no basis in fact whatsoever. There never was a time when fine-tuning investigations varied just one parameter.

The original anthropic principle paper by Brandon Carter in 1974 identified a peculiar relationship between the mass of the proton, the mass of the electron, the strength of gravity and the strength of electromagnetism.[11] Stars can transport energy from their nuclear-burning cores to their surface in two different ways – in the form of radiation, or via *convective currents* in which warmer gas rises and colder gas falls in cycles. In universes that subscribe to Carter's coincidence, both kinds of stars are possible. Carter conjectured that life requires both kinds for heavy element production and planet formation.

---

[11] If you're a physicist or mathematically inclined, enjoy:

$$\left[\frac{G}{\hbar c} \frac{m_{proton}^6}{m_{electron}^4} \right]^{\frac{1}{8}} \sim \alpha^{\frac{3}{2}}$$

There is no known reason for any of these numbers to be related to any other, and yet the coincidence holds to within 15%: $7.1 \times 10^{-4} \approx 6.2 \times 10^{-4}$.

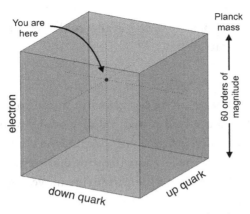

FIGURE 40 A three-dimensional Etch a Sketch®, representing 'parameter space'. To help visualize the possible values of the masses of the fundamental particles, we imagine choosing a point in the block. As we spin the dials and choose different masses, our stylus moves through the block. Where can life flourish?

Physicists William Press and Alan Lightman showed in 1983 that the same coincidence must hold for stars to emit photons with the right energy to power chemical reactions. This is quite a coincidence, given the number of cosmic dials one must tune for the energy of a photon of light emerging from a star to be roughly equal to the energy of chemical bonds.

The whole point of this relation and many more like them, with which the early anthropic literature is entirely concerned, is that they relate *a number of different fundamental constants.*

More recent work has shown that spinning multiple dials is usually as destructive as spinning one. Suppose we spin the up quark, down quark and electron mass dials. Recall that you are made of these three particles: two up quarks and one down quark make a proton, and one up quark and two down quarks make a neutron.

Think of Figure 40 as a three-dimensional Etch a Sketch®, where one dial sets the up quark mass, one sets the down quark mass and the other sets the electron mass. When you've dialled in the masses, the stylus is at a particular point in the block. When physicists talk of 'parameter space', this is something like what we have in mind.

What are the ranges of our dials? Or, to put it another way, how wide is our block? On the lower end, particles can have zero mass – the photon, for example. What about the upper end? Here things are trickier, but there is at least a firm edge to our knowledge. At the moment, we don't have a theory of quantum gravity. That is, we don't know what happens when quantum things (such as particles) are gripped by their own gravity (as in a black hole). A simple calculation suggests that a particle with a mass equal to the *Planck mass* would become its own black hole. So this is the maximum mass that our theories – absent quantum gravity – could possibly handle. The Planck mass is roughly 24,000,000,000,000,000,000,000 (2.4×10^{22}) times the mass of the electron!

This mass is so large that, to help illustrate the interesting bits in our block, we need to use a *logarithmic scale*. It's an easy idea: instead of each click of the dial moving the masses in the usual 0, 1, 2, 3 … manner, we instead multiply by ten: …,.0.01, 0.1, 1, 10, 100, …

Stephen Barr and Almas Khan of the University of Delaware (2007) took the stylus to all corners of a two-dimensional slice through our block to investigate changes in the quark masses. We'll go one better by investigating the three-dimensional block. In the specific model they investigated, their lower mass limit is set by something called 'dynamical breaking of chiral symmetry' to be about 60 orders of magnitude – 10^{60} – smaller than the Planck mass. We will do the same for each side of our block.

To help our novice universe builder to avoid disaster, we'll carve off parts of the block that we have identified in previous chapters as being unsuitable for life; we met many of these universes in Chapter 2. For example, in Figure 41 we've carved off the disastrous Delta-plus-

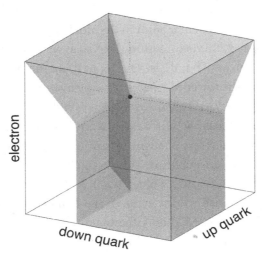

FIGURE 41 Carving off failed universes, Stage 1. Starting with the block of Figure 40, we carve off the Delta-plus-plus, Delta-minus, hydrogen-only and neutron-only universes, in which there is at most one chemical element and one possible chemical reaction.

plus universe, with one stable element and no chemical reactions, as well as the simply appalling Delta-minus universe, with one element and one chemical reaction. In fact, we'll go a step further by carving off the hydrogen-only universe and the 'worst universe so far', the neutron universe – no elements, no chemistry.

Stable atoms have a few more regions to avoid. We'll carve off the parts where protons and neutrons don't stick to create nuclei. We'll carve off the regions where the electron can be captured by the nucleus, reducing atoms to piles of neutrons. We'll carve off the parts where anything with the chemistry of hydrogen is unstable. Figure 42 shows what remains.

Further, our dials are messing with stars' nuclear fuel and the source of their internal pressure. We'll carve off the region with no stable stars at all, as identified by Fred Adams. We'll also ensure that the first product of stellar burning (the *deuteron*) is stable, and that

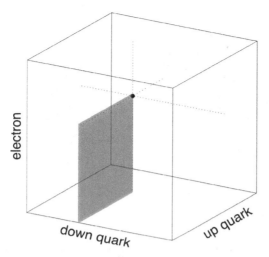

FIGURE 42 Carving off failed universes, Stage 2. We remove regions of the block where atomic nuclei fail to be stable at all.

its production releases energy rather than absorbing it, since this would upset the gravity vs. thermal energy balance in a star. Figure 43 shows what survives our slicing and dicing.[12]

Finally, we carve off universes in which the Hoyle resonance (Chapter 4) fails to allow stars to produce both carbon and oxygen,[13] which leads to Figure 44.

What remains is a thin shaft of life-permitting universes extending to small values of the up quark mass, surrounded by a vast wasteland. Remember that we needed to use a logarithmic scale; we can now see why. If we used a normal (*linear*) scale from zero to the Planck

[12] As explained in Chapter 4, unlike Barr and Khan, we don't carve off universes in which the diproton is bound. Contrary to some claims in the fine-tuning literature, binding the diproton does not necessarily cause all the hydrogen in an early universe to be burned up. Further, stars that burn by first forming the diproton can be stable and long-lived. They are not necessarily explosive.

[13] Epelbaum et al. (2013) show that changing the sum of the light quarks by 2–3% is enough to affect carbon and oxygen production. In our plot, we show a generous 20% change.

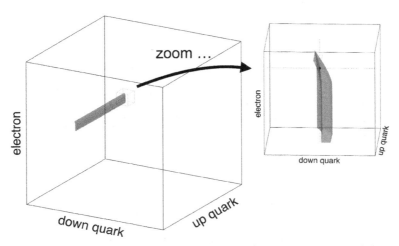

FIGURE 43 Carving off failed universes, Stage 3. If a universe fails to support stable stars, then it is cut out of our block.

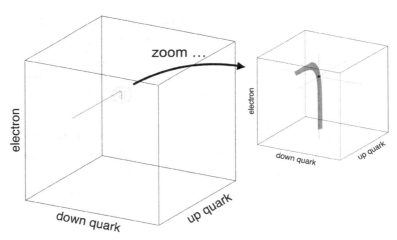

FIGURE 44 Carving off failed universes, Stage 4. As we discussed in Chapter 4, a special property of carbon nuclei (the Hoyle resonance) allows stars in our Universe to make both carbon and oxygen. We remove from the block universes in which this fails.

mass, we would need a block at least 10 light years (a hundred billion kilometres) high for the life-permitting region visible to the human eye.

The problem with this reaction is obvious. Sure, there are many dials. But there are also many requirements for life. Adding more dials opens up more space, but most of this space is dead. We see no trace whatsoever of a vast oasis of life.

Life is similarly confined in cosmological parameter space. Max Tegmark, Anthony Aguirre, Martin Rees and Frank Wilczek (2001) find eight constraints on seven dials. Again, life is left to huddle in a tiny island. (Wilczek is a Nobel Prize-winning particle physicist, and Rees is the Astronomer Royal and former president of the Royal Society.) We'd love to plot all seven dimensions for you – blasted two-dimensional paper!

This myth may have started because, when fine-tuning is presented to lay audiences, it is often *illustrated* by describing what happens when one parameter is varied. Martin Rees, for example, does this in his excellent book *Just Six Numbers*. Rees knows that the equations of fine-tuning involve more than one parameter – he derived many of those equations.[14]

Two fallacies must be avoided. The first is focusing on the shape of the life-permitting island, rather than its size. As we saw above, the life-permitting island is not a single blob. In general, it could snake through the dimensions of parameter space. We could say that life is possible for a range of values, but this would be misleading. We still need to carefully adjust the dials. A random spin of each dial is unlikely to result in success.

---

[14] For example, from Tegmark and Rees (1998):

$$\alpha^{-1}\ln(\alpha^{-2})^{-16/9}\alpha_G\left(\frac{\beta}{\zeta}\right)^{4/3}\Omega_b^{-2/3} \lesssim Q \lesssim \alpha^{16/7}\alpha_G^{4/7}\beta^{12/7},$$

which obviously involves more than one variable, but is a tad too intimidating for a popular audience. Unless you hide it in a footnote. Also note that this is an inequality, not an equation, and so should not hurt sales of our book.

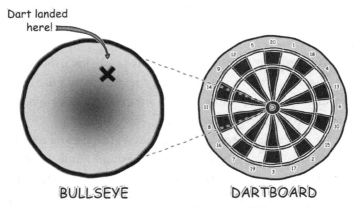

FIGURE 45 A dart lands inside the bullseye. It could have landed twice as far away from the centre and still scored a bullseye. Does that mean that the throw was only 'fine-tuned' to within a factor of 2, or that scoring a bullseye was a fifty-fifty chance? Obviously not! The smallness of the bullseye compared to the size of the wall – the set of places that the dart could have landed – could be evidence of dart-throwing prowess.

The second fallacy is comparing the life-permitting range to the value of the constant in our Universe. Here's an analogy. Suppose you throw a dart at a board and it lands inside the bullseye, 3 mm from exact centre (see Figure 45). Not bad, eh? Not so fast, says your friend. You could have landed twice as far from the centre and still scored a bullseye. So your throw is only 'fine-tuned' within a factor of 2 ... not very impressive at all!

Something has gone wrong here. It is the size of the bullseye compared to the size of the wall – not compared to where your dart landed – that makes a bullseye evidence of either your dart-throwing prowess, or your determination (despite your terrible aim) to keep throwing until you hit the bullseye.

We noted in Chapter 2 that increasing the mass of the down quark by a factor of 6 results in the atom-, chemistry-, star- and planet-free neutron universe. This might seem like plenty of room. While 'a factor of 6' is fine for *stating* the limits of the fine-tuning region, it gives a misleading impression of its *size*. Compared to the highest energies that particle accelerators have reached, the life-permitting range is less than one in a hundred thousand. Compared to the Planck mass, it is one part in 10^{20}. The range of *possible* values of a constant (in a given theory) is often far larger than the *actual* value.

Reaction (j): Life Chauvinism – Why Think That Life is Special?
The comment: You're so vain! You probably think this Universe is about you, don't you?[15] There are loads of kinds of things that only exist in a small fraction of possible universes. Why aren't we discussing fine-tuning for planets, or black holes, or iPads? Like the creation myths, fine-tuning simply assumes that humankind is the most special thing in the Universe. It's self-importance on a cosmic scale. The Universe doesn't care about us.

The short answer: A fact about our Universe can be thought of as *special* if it supports some theories but not others, since then it helps us test our ideas.

The long answer: Is there anything special about the way the Earth's continents seem to fit into each other like a jigsaw puzzle? Is there anything special about Brownian motion, the random jostling of particles suspended in a liquid? Is there anything special about the faint noise that physicists Arno Penzias and Robert Wilson couldn't quite eliminate from their antenna?

While these observations seem inconsequential, they are very special because they support important theories about how the world works. Alfred Wegener showed that the continental jigsaw is evidence that they were once joined, and have moved apart on tectonic plates. Einstein showed that Brownian motion is evidence of the existence of atoms, measuring their tiny masses. And it was only when Penzias and Wilson's experiment came to the attention of Robert Dicke and his colleagues at Princeton University that it was realized that they had

---

[15] With apologies to Carly Simon.

discovered the cosmic microwave background, a relic of the early Universe.

Here's another example. Are the striations – the pattern of groves and scratches – on a bullet at a crime scene special? Well, you might think, *all* bullets that have been fired must have *some* pattern, so how could this one be special? However, the striations suddenly become very important if we think we've got the gun that fired the bullet, since they are like a fingerprint. The striations are special to the theory that *this* gun fired the bullet.

Facts can be special *to a theory*. They are special because of what they tell us about the world, because of what we can *infer* from them.

Is life special? Well, it's certainly a scientific fact that this Universe contains life. And *any* scientific fact could turn out to be special – it depends on what theories we have about the Universe. We'll get to these soon enough, but first we can note some *prima facie* evidence that life could, in this sense, be special.

'If I didn't have you', sings comedian Tim Minchin, 'someone else would do'. His rather blunt love song continues: 'I mean, I think you're special. But you fall within a bell curve.'[16]

When looking for something special, finding something *rare or unusual* is at least a start. Minchin's beloved is, frankly, replaceable. There are a large number of other possible partners and 'statistically, some of them would be equally nice'.

Life, on the other hand, is very unusual in the set of possible universes, as we have seen. This is in stark contrast to, for example, black holes. Those are easy – just let gravity do its thing. As we saw in Chapter 5, increasingly the lumpiness (Q) of a universe results in black holes galore.

But rareness is not enough – iPads are at least as unusual as life, and no one's written a book about the fine-tuning of the Universe for iPads. Why would life be special to a theory, rather than iPads? Here are two ways.

---

[16] From the album *Ready for This!* (2009), used with permission.

As observers: Life forms may be special to science, because scientists are life forms.

In science, what we observe depends not only on what is out there, but also on *with what* we are observing. What you see in the night sky depends an awful lot on what you're looking through. It depends on the atmosphere, the mirror, the detector, the software ... and the sentient meat sitting in front of the computer screen.

For example, in any survey of the distant universe, we only detect the brightest members of the population. We can't see what is too faint to see. Such *selection effects* are what keep astronomers up at night.[17]

Life is special because it comes attached to a selection effect. As we noted in Chapter 1, the Universe is not our experiment. We are not Dr Frankenstein, we are the monster. We are a product of the very thing that we are studying.

Science has worked very hard to remove, or at least compensate for, the effect of human beings and their loveable foibles and biases. We use double-blind medical trials to overcome the placebo effect. Part of the attraction of using mathematics to formulate theories and analyse data is that it can be made automatic and algorithmic. We can get a computer to do it, removing the human almost completely. We get other experts to review our work before it is published. We control variables, calibrate instruments, and collaborate with other experts.

But at the end of it all, science is a type of knowledge, and so science needs a knower. What makes science possible is your brain, the most complicated kilogram-and-a-half of matter in the Universe. Not all laws of *nature* can become laws of *science* because many will not create a scientist.

As moral agents: Life could be special because life forms are capable of moral knowledge and moral actions.

---

[17] Get it?

We are insignificant, we are told, because we are so small and the Universe is so big. It is well known, and you have seen it demonstrated by astronomers, that beside the extent of the heavens, the circumference of the Earth has the size of a point; that is to say, compared to the magnitude of the celestial sphere, it may be thought of as having no size at all. This puts a dent in medieval ideas of human specialness ... at least, it would, but for the fact that the previous sentence is a quote from the sixth-century philosopher Boethius. Earth is a cosmic speck; this is not new information. As we noted in Chapter 1, preconceived ideas of human self-importance have actually had a rather small effect on theories about the Universe.

Keep in mind that *significance, purpose, relevance, importance, specialness* ... these are not scientific terms. We haven't invented a significance-oscope. No purpose-ometer. There is no term in our equations for relevance. We haven't surveyed the Universe with a telescope that measures importance. Science cannot state that we are insignificant.

To call a person *significant* is to acknowledge their *moral* worth. This is a category that isn't measured by telescopes or in metres. We are creatures that can investigate evidence, seek truth, reason logically, create music, admire beauty, stand in awe, treasure goodness, admire virtue, and demonstrate love. *Persons* measure importance, and as Dr Seuss says, 'a person's a person, no matter how small!'[18]

Reaction (k): We Don't Even Have a Good Definition of Life

The comment: 'Life as we know it' isn't particularly well understood, especially when it comes to the conditions under which life develops from non-living chemicals. Life has proven remarkably difficult to define.

The short answer: We discussed this reaction in the first chapter, concluding that the carnage awaiting life in parameter space

---

[18] From *Horton Hears a Who!* (1954).

makes worries about the different definitions of life into a minor technicality.

Reaction (1): There Could Be Other Forms of Life

The comment: To properly establish the fine-tuning of the Universe for life, we need to calculate the probability of *every kind* of life. What about forms of life that we just haven't considered? For example, you've discussed the conditions under which carbon is stable and forms inside stars. But what about silicon-based life? What about life based on other chemicals? What about life not based on chemicals at all?

The short answer: The discovery of other possible forms of life is not enough to overturn fine-tuning. We would need to discover alternative forms of life that could exist in any old universe, including the vast wasteland that we discarded as uninhabitable. No such forms of life have been proposed. Furthermore, if life is so easy, why does life as we know it rely on such extraordinarily complicated organic chemistry?

The long answer: Firstly, most of the fine-tuning cases discussed above have assumed very little about life. Consider the cosmological constant: a seemingly small increase led to a universe with no structure whatsoever, and a small decrease (to negative values) leads to no universe whatsoever. These universes are so simple that there is nowhere for alternative life forms to hide.

Secondly, even if *some* life form could escape the carnage we have discovered in parameter space, it is still the case that seemingly small changes in the free parameters of the laws of nature as we know them have *dramatic, uncompensated* and *detrimental* effects on the ability of the Universe to support the complexity needed for physical life forms. Even if we cannot *with certainty* rule out the existence of life in such universes, there is still something noteworthy about the suitability of our Universe for life.

Thirdly, where we have focused on carbon-based life forms, it is not just 'carbon chauvinism', that is, a fondness for our own biochemistry. Carbon, it turns out, is rather unique.

Consider carbon's siblings on the first row of the periodic table of elements: lithium (Li), beryllium (Be), boron (B), nitrogen (N), oxygen (O), fluorine (F) and neon (Ne). Now ask: how many chemical compounds can be made using each element and hydrogen, the simplest element of them all?

Li	Be	B	C	N	O	F	Ne
4	6	38	29,019	65	21	6	0

Hopefully, one of those numbers strikes you as unusually large.[19] You can make an awful lot of things with carbon. It is versatile and flexible in a way that is unmatched by any other element. Remember, there are only a finite number of elements. We can test them all.

Carbon's flexibility gives it the ability to make information-storing molecules. These molecules can be *big*, since carbon can form long chains; a lot of information is needed to specify the recipe for you. Carbon compounds are *metastable*, that is, they are not so stable that the information cannot be read and utilized, but they aren't unstable either – they must be induced to interact, and so do not spontaneously dissolve their information.

Further, the great variety of carbon's compounds makes DNA an efficient message medium. The chemistry of life allows the letters of the DNA alphabet (usually denoted G, A, T, C) to be arranged in any order, without any chemical affinity between the letters. This is essential, for the following reason. Suppose your computer keyboard malfunctioned, so that whenever you pressed 'A' it would add a 'C'. This would severely compromise your ability

[19] These numbers come from a search of the chemspider.com database, September 2015. The table includes chemical compounds with net electric charge, and disregards isotopically labelled structures.

to type meaningful messages. Similarly, if DNA's G attracted A more than C, then the DNA message would contain redundancies, sequences that were due to mere chemical attraction rather than the message being codified. Carbon's versatile chemistry does not include such affinities, and so can truly underwrite a code.

Water, too, distinguishes itself from the pack. Many of its properties are anomalously different to comparable chemical compounds. For example, it can efficiently absorb heat; as a result, Earth's oceans play a significant role in stabilizing our climate. Water is a near-universal solvent, and so can be used efficiently by life forms for internal transport. Carbon dioxide (CO_2), a common waste product of metabolism, is a gas that dissolves in water. Its disposal is relatively easy: just pipe it through your veins to the lungs and breathe it out.

Further, water's solid form is less dense than its liquid form; or, more familiarly, ice floats. Without this property, ice would not form a thin, insulating surface layer on oceans and lakes; rather, they would freeze from the bottom up. An ice age could freeze the oceans solid, wiping out all aquatic life and leaving the Earth a frozen snowball. Ice effectively reflects sunlight, so the snowball could be permanent. (Remember: the point is not that water-less life is impossible, but that it is probably more difficult to form.)

Now consider carbon's closest sibling, silicon, which is directly below carbon in the periodic table. You may remember from high-school chemistry that elements in the same column of the periodic table have similar chemical properties. Silicon, like carbon, has four outer electrons available for chemical bonding. But their similarities pale in comparison with their differences.

Silicon can form 55 compounds with hydrogen; this is dwarfed by carbon's 29,019 (as we saw above). While silicon can form long chains, these tend to be repeating. In particular, the equivalent of CO_2 for silicon-based life – you guessed it, SiO_2 – is a crystal. Instead of a water-soluble gas being formed in every cell of your body, you'd have sand.

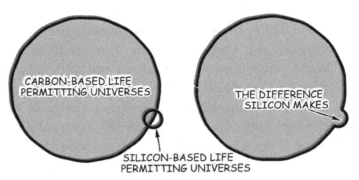

FIGURE 46 A comparison of the life permitting universes of silicon and
carbon, showing that silicon, while it possesses the potential to provide
the framework for the chemistry of life, is a mere pimple compared to the
potential of carbon.

This isn't a deal breaker for life – that's not the point. We can
perhaps imagine a form of life that overcomes silicon's deficien-
cies. Rather, we're trying to get a bird's eye view of life in parameter space. If silicon-based life requires even more special
circumstances than carbon, then its patch in parameter space is
probably smaller. Certainly, silicon-based life won't exist in just
any old universe.

Furthermore, silicon and carbon are both synthesized in
stars and supernovae. Recall that stars make large nuclei by first
making small nuclei and then smashing them together. So, any
universe that makes silicon will probably first make carbon.
In that case, the silicon-based life patch overlaps with the carbon-
based life patch. In our Universe, silicon is not in short supply:
there is 150 times more silicon than carbon in the Earth's crust.

The situation with silicon is illustrated in Figure 46. The net
effect of including life based on silicon is a small pimple on
carbon's already pint-sized piece of parameter space. Silicon-
based life obviously does not fill a substantial portion of parameter
space.

Let's broaden our search, let our imagination run wild ... and consult the impressively detailed Wikipedia page: 'List of fictional extraterrestrials by form'. From Asgardians to Zygons, and from The Blob to The Thing, they're all there.[20]

Many of these aliens simply will not do. Some are just variations on life as we know it; Chewbacca, while adorable, is a combination of a human, a dog, and a crossbow. These represent not alternative forms of life, but rather explore the contingencies of evolution. Perhaps a left turn during the Cambrian would have replaced humans with Xenomorphs, Ewoks, Decapodians or Gorns.[21] These aren't the aliens we're looking for.

Other science fiction writers simply paste a human-like mind into an unusual place, like a gas cloud, a pillar of light (or 'pure energy', whatever that is), or a killer tomato. While this will do just fine to advance the story, we're missing a few details. How, exactly, do you arrange for a gas cloud to process and store information? There's no physics here, and so no way to search parameter space for sentient gas clouds.

If we try to fill in the details, things get even worse. Suppose that, inside our sentient gas cloud, we arrange for collisions between the particles to perform calculations and store information. Computer scientists have worked out how to simulate a computer using billiard balls; recently, a similar design was implemented using soldier crabs. We have no idea why.

We program the initial state so that, once all the collisions are complete, the final state of the particles records the answer. To predict

---

[20] One of us – do your own detective work – claims that, by way of descending from the *Silures*, a Celtic tribe of Wales, he has a direct lineage with the *Silurians*, a race of reptile-like humanoids who share a universe with Dr Who. The other author's ancestors are disappointingly non-fictional.

[21] Congratulations if you got all of those. We'll save the rest of you some Googling. The alien in *Alien* and *Aliens* is a Xenomorph. Ewoks are the Empire-defeating teddy bears from *Return of the Jedi. Futurama's* Dr John Zoidberg is a Decapodian. Finally, Star Trek's Captain James T. Kirk battled a Gorn, a reptilian biped known for its rubbery skin, emotionless face and slow-motion fighting style.

the paths of all the particles, we'll need to take into account all the internal and external forces.

Here's the problem. The billiard ball computer needs the balls to bounce in a precise direction and maintain their path; this makes external forces irrelevant. It needs a solid structure with rigid walls and rubberized cushions. In the gas phase, we have no such luxury.

Collisions between gas particles are *chaotic*. The exact path that a particle takes from collision to collision depends very sensitively on every minute detail of the system and its environment. The more collisions you need to calculate, the more precisely you need to know the details of the cloud and its surroundings. In 1978, Michael Berry showed that in order to predict the next 56 collisions of a molecule of oxygen in this room, one must know the position of every particle in the observable Universe! Ignoring the miniscule gravitational pull of a single electron 10 billion light years away generates errors that render the prediction useless in about 10 nanoseconds.

Fred Hoyle's 1957 novel *The Black Cloud* (spoiler alert!) tells of the discovery of an enormous cloud of gas that is headed for the Solar System. The cloud turns out to be a super-organism, a sentient creature (nicknamed *Joe*) that is vastly more intelligent than human beings. Joe is somewhat surprised to find life on a planet.

Hoyle's marvellous book is highly recommended, and (of course) does an excellent job in presenting a vivid portrait of the science and scientists.[22] However, Joe will face the chaos problem. Your brain sends signals along solid (if flexible) neurons, ensuring that they get to their intended destination. Sending a particle through the gas phase, on the other hand, is fraught. Repeating Professor Berry's calculation, it will take only 11 collisions – 1/1,000 of a second – for

---

[22] Hoyle, a theoretician, could be impatient with his observational counterparts. In his novel, he describes a talk by an amateur astronomer: ' . . . Mr Green suddenly seemed to remember the purpose of his talk. Quitting the description of his beloved equipment, he began to throw off his results, rather like a dog shaking itself after a bath.' (1957, p. 31).

a particle to be thrown off course by an electron at the edge of the Universe.

Joe overcomes this problem, we are told, using magnetic fields as if they were giant arms, and by sending particles along large-scale fluid flows. These magnetic fields also prevent the cloud from collapsing under its own gravity. This is all controlled from a central brain, constructed from organic molecules. Joe roams the universe, occasionally visiting a star to repower and replenish its chemical stores. Sentient clouds are born when one of Joe's kind discovers a suitably organic material-rich interstellar cloud and seeds it with life.

In the book, a scientist remarks: 'The thing which staggers me is the astonishing similarity in the principles on which life is maintained. ... The details are of course wildly different: gas instead of blood, electromagnetic heart and kidneys, and so on. But the logic of the layout is the same.' (1957, p. 180)

If we went searching for Joe in parameter space, we would find that his universes of choice are very similar to ours. Joe's requirements – stable nuclei and atoms, interesting chemistry, long-lived universe, long-lived and chemistry-powering stars, supernovae to distribute chemical elements into interstellar space, galaxies to collect matter into clouds and hence stars – are familiar. Joe, too, is living in a fine-tuned universe.

The problem with most fictional alien life forms is that we aren't told how they work. Hoyle's Joe is a wonderful exception, and a case in which the inner workings are similar to our own. Life as we know it has an extraordinarily complex array of processes working behind the scenes. The cell is not a homogeneous blob of protoplasm; indeed, nothing could be further from the truth. The cell can be thought of as a colossal factory, a city even, a Rube Goldberg contraption of interconnected assembly lines, machines, highways, transporters, blueprints, and more. It would be extremely surprising if life on Earth went to all that bother if life is easy to make.

Consider also cellular automata, discussed in the previous chapter. Conway's game is about as different from our Universe as we could hope, and yet its rules are fine-tuned. Interesting rules are a rare commodity in the set of possible rules.

Hoyle himself commented in his 1950 BBC Radio broadcast 'Man's Place in the Expanding Universe':

> It is my view that man's unguided imagination could never have chanced on such a structure as I have put before you. ... No literary genius could have invented a story one-hundredth part as fantastic as the sober facts that have been unearthed by astronomical science. You need only compare our inquiry into the nature of the Universe with the tales of such acknowledged masters as Jules Verne and H. G. Wells to see that fact outweighs fiction by an enormous margin.[23]

The real Universe is surprising, something you could not have guessed. Life is no exception. We cannot, of course, take into account things we haven't thought of. But that is always true, and in this case there is no reason to think that what we don't know will overturn what we do know. As we have learned more about life and its place in the Universe, the case for fine-tuning has become stronger. More importantly, the reason that the Universe couldn't have been guessed is that it is more interconnected, more complex, more intricate than we could imagine.

But a fine-tuning-busting life form would have to be simple, if it is going to inhabit the vast wastelands of parameter space, with scarcely a molecule for company. The absence of any such candidate is a loud silence indeed. And if life could be that easy, why are we so complicated?

Reaction (m): The Anthropic Principle Explains Our Existence

The comment: What else would we expect from a universe that contains us, but that it is able to evolve and support us? We are here, so

---

[23] Published in *The Nature of the Universe* (1950, p. 118).

obviously we can be here. How can we be surprised by an observation that follows necessarily from the fact that we are observing at all?

The short answer: Why do life forms exist at all? There are plenty of possible universes in which no observations ever happen. Why does our Universe exist, rather than one of those? Facts that we deduce from our existence do not explain why we exist at all. The conditional statement '*if* physical observers *then* an observer-permitting universe' does not answer the question: 'why observers?'

The long answer: We can formulate the argument as follows.

(1) Necessarily, if physical observers exist, then the Universe permits the existence of physical observers.
(2) Physical observers exist.
(3) Thus, the Universe necessarily permits the existence of physical observers.

Despite its seemingly logical form, this argument is invalid. The fallacy is exposed with an example.

(A) Necessarily, if I have four apples, then I have more than two apples.
(B) I have four apples.
(C) Thus, I necessarily have more than two apples.

From (A) and (B), it is true that I have more than two apples. But this does not mean that it was impossible for me to have had no apples; it is not *necessarily* true. The rules of arithmetic (A) don't give out apples.

Modern *modal* logic expresses this idea in terms of *possible worlds*. As we will explain in more detail in the next Reaction, a possible world is a *total* way that reality could be, a consistent state of the world to which no further detail could be added without rendering it inconsistent. A statement is necessarily true if it is true in all possible worlds. So, we can rewrite our premises as follows. (A') In all possible worlds in which I have four apples, I have more than two apples. (B') In the actual world (i.e. this world), I have four apples. From this, it follows that I have more than two apples *in the actual world*, but it does not follow that I have more than two apples *in all possible worlds*.

Note well: (A) remains true even if I have no apples. Similarly, even in a universe that contained no life and no observers, it would still be true that 'Necessarily, *if* physical observers exist, *then* the Universe permits the existence of observers.' Clearly, (A) cannot prove the existence of apples, and (1) does not make life inevitable.

This reaction to the fine-tuning of the Universe for life is not found in the scientific literature, probably because scientists – and astronomers in particular – understand *selection effects*.

We met selection effects above; let's have a closer look. Suppose you want to know who will win the next election. Ideally, you would survey everybody who can vote.[24] However, this is impractical, and you will save a lot of time and money if you survey a *representative* sample: a large enough, varied enough group of people should let you infer how the population as a whole will vote.

So you survey 100 people down at your local shopping centre. How can you test whether your *sample* is representative of the *population*? You could, for example, record the ages of each survey participant. Unbeknownst to you, it was pension day: 20 per cent of your sample are over 65, in contrast to 10 per cent of the general population. Your sample isn't useless; to compensate, make each participant represent a different fraction of the population, depending on their age. But you could reach the wrong conclusion if you treat your sample as if it were the population.

Behold, the selection effect: between the sample (what you have) and the population (what you want to know about) is the process of selection. Ignore at your peril.

In astronomy, what we observe doesn't just depend on what you're looking *at*; it also depends on what you're looking *with*. Suppose your survey of galaxies reveals that, the further away the

---

[24] Trivia: in Australia, voting is compulsory; non-attendance incurs a fine. Geraint was fined $50 for missing a council election a matter of months after becoming an Australian citizen.

galaxy is, the more stars it has. Does this mean that more distant galaxies have more stars?

Wait a minute! We just jumped from a statement about a *sample* of galaxies to a statement about the whole *population* of galaxies. Let's think for a minute about our sample selection. We pointed our telescope at the sky, looking for bright *blobs* of light. (Bright *points* of light are probably stars in our galaxy.) To be seen, a galaxy must be bright enough to poke its head above the faint glow of even the darkest sky, seen with the biggest telescope. The more distant the galaxy, the more *intrinsically* bright it will need to be in order to be seen by us, in the same way that you can see a candle across the room, but it takes a lighthouse to be seen across a lake. And so the further we look, the more stars a galaxy needs in order to be selected in our sample.

Thus, the relationship between distance and average number of stars per galaxy in our *sample* can be explained without supposing that there is a relationship between distance and number of stars in the *Universe*.

Selection effects, by themselves, explain nothing. They connect a population with a sample. The answer to the question 'why are galaxies so bright?' is not 'because otherwise we wouldn't see them'. There must first *be* a population that includes bright galaxies in order for any of them to be in our sample.

'Population + selection effect'-style explanations are common in science. For example, *publication bias* is the tendency of boring, 'we didn't find anything significant' results to remain in a researcher's desk drawer. But if only the experiments that found a positive effect of MiracleDrug™ are reported, MiracleDrug™ will seem better than it is.[25] Such effects are very important, but go nowhere without a population. Selection effects *alone* explain nothing.

---

[25] We heartily recommend *Bad Science* (2009) and *Bad Pharma* (2014) by Ben Goldacre, books that highlight the impact of selection biases on science in modern society, especially the reporting of medical breakthroughs.

We conclude that the anthropic principle – that observers must inhabit a universe that permits observers – is powerless on its own. It explains why we *don't* observe a life-prohibiting universe, but it doesn't explain why a life-permitting one exists at all. However, if there were a population of universes, many and varied . . . more of that in the next chapter.

Reaction (n): Whence the Possibilities?
The comment: How do we know that all these other universes really are possible? We have no idea how constants are set in these universes, nor the possible ranges of the constants, nor even the possible kinds of universes there might be.

The short answer: The other universes that we have considered are free from internal contradiction, since we have changed the laws of nature in ways that do not affect their mathematical consistency. If you believe that there is some stronger principle that dictates what is possible and impossible, that for some reason disqualifies certain mathematically consistent universes, then define it, defend it, and explain why it is so fond of stars, planets, chemistry and life. Those are exactly the kinds of explanations we're looking for!

The long answer: 'Possible' and 'impossible', along with 'necessary' and 'contingent' (neither necessarily true nor necessarily false), are terms that have been the subject of a few millennia of philosophical debate. We have no desire (or indeed, ability) to attempt here a lecture on *modal* logic.

Three senses of 'possible' could be relevant to fine-tuning.

Absolute possibility: The most obvious way that some state of affairs could be impossible is by being self-contradictory. A universe that both does and does not contain electrons is a non-starter. Recent philosophical thought has discussed a broader idea: absolute (or *metaphysical*) possibility. There may be ways that the world cannot be, even though we cannot find a strict logical contradiction in its specification. For example, it doesn't seem possible that you could have been a crocodile. There is no contradiction in that statement, but anything that is a crocodile is not you.[26] Similarly, the prime minister could not have been a prime number.

Now, consider the kind of claims we've been making thus far: if the laws/constants of nature had been A, then the universe would have behaved like B. These *counterfactual conditionals* are familiar, yet may seem strange on closer inspection. By definition they start with a false (counter-to-fact) statement, e.g. if we had not written this book, you would not be reading it. Physical laws are often thought to imply counterfactuals: if two masses were separated by a certain distance, then the mutual force of attraction between them *would be* given by Newton's law of gravitation.[27]

Counterfactual conditionals can be analysed in terms of (absolutely) *possible worlds*. A possible world is a *total* way things could be, that is, a consistent state of the world to which no further detail could be added without rendering it inconsistent. So, consider the counterfactual statement 'if you had been late for work, you would have been fired'. One way of understanding this claim is as follows: among the possible worlds in which you are late for work, the ones in which you are also fired are closer (or more similar) to the actual world than the ones in which you are not fired. For example, being fired requires only that your boss maintain his tyrannical ways; keeping your job would require uncharacteristic mercy.

So – getting to the point – are the other universes of fine-tuning *absolutely* possible?

Here, fine-tuning seems to be on safe ground. We begin with the laws that describe the actual universe, whose mathematical self-consistency is well studied. We then change the values of arbitrary constants, upon which the consistency of the equations does not

---

[26] Perhaps (following Chalmers, 1996) this kind of metaphysical possibility is a constraint on the statements we use to describe possible worlds, rather than on possible worlds themselves.

[27] In this case, the antecedent (that two masses are separated by a certain distance) isn't necessarily false; such statements are sometimes called 'subjunctive conditionals' to distinguish them from the narrower sense of counterfactual, which involve a false antecedent. We will use the broader sense of *counterfactual* to cover both cases.

depend. We are about as close as we could hope to a mathematically consistent and complete description of another possible world.

Physical possibility: If mathematical consistency gives our alternative universes a passing grade, suggesting that they are absolutely possible, what about a deeper *physical* principle? Are these universes physically possible?

Recall from Chapter 1 that physical theories have four components: the stuff, the constants, the dynamics (described by a mathematical law) and the situation. Given the first three of these parts, we can assemble the set of mathematical solutions to the laws. These solutions represent situations that are consistent with the law, that is, universes that the law declares possible. These are the physical possibilities of a theory.

Different theories involve different physical possibilities. Faster-than-light spaceflight is physically possible according to Newtonian physics, but physically impossible according to Einstein's Special Relativity. So, are the other universes of fine-tuning physically possible?

When we alter the initial conditions of a theory, we are considering different *physically possible* universes. The fine-tuning of initial conditions, then, explores exactly the set of possibilities that the laws of nature have provided. So that's on safe ground.

What about if we change the constants in the laws, or even the laws themselves? If we postulate that the laws and constants in question are part of *the ultimate* laws of nature, then by definition there are no deeper laws to stand over our proposed laws and constants, disqualifying them as physically impossible. They make the rules of the game. As an analogy, it cannot be against the laws of chess to invent the game of checkers. Thus, changing these laws and constants is above any considerations of what is and isn't physically possible.

If, on the other hand, the laws whose constants we vary are approximations of deeper laws, then it is an open question whether changes to the 'constants' that we know are physically impossible (if they are really mathematical constants) or physically possible (if they are dynamical entities such as fields, or can be written in terms of other constants). But this is exactly the question that fine-tuning should lead us to ask, and might be a clue as to its answer – more in the next chapter.

Whatever the case, we are still left with the question: why are some absolutely possible worlds singled out as the *physically* possible worlds? Understanding the consequences of different laws can provide a clue to answering this question.

Conceptual possibility: Before we start speculating on what is absolutely possible or ultimately physically possible, we should take stock of what we know. We should consider what is *conceptually possible*, that is, *for-all-we-know* possible.

Since the constants and initial conditions that we have varied in this book so far are found in the deepest laws of nature that science has discovered, nothing that we *know* rules out these other universes. They are, *for all we know*, possible.

But suppose that there is, unbeknownst to us, some deeper logical, metaphysical or physical principle that strongly constricts the set of possible universes. We may find that many of the universes we have investigated are, in fact, ruled out. Does this paralyse our investigation? Should we simply stop until we can be certain that no such deeper principle exists? Not at all! Given a set of ideas about how the world works, the thing to do is test them. A deep principle of this sort is another idea to test.

In short, no sense of *possible* seems to create problems for fine-tuning claims, and all we really need is the least problematic of them all: conceptual possibilities.

If the other universes of fine-tuning turn out to be impossible, it will be because a powerful, deep, as-yet-unknown principle of ultimate reality has shown a firm and fortunate fondness for micro-managing cosmic minutiae, from the amount of matter to the electron's mass. We want to know what that principle might be, and whether it is actually true. (We might even dare to ask whether it is *necessarily* true.) Our curiosity has failed us if we ignore for-all-we-know possible universes without good reason.

Reaction (o): Whence the Probabilities?

The comment: OK, so I'll grant you some possible universes. But how are you going to assign probabilities to such an unwieldy mob? You've got more universes than you can poke a stick at. Furthermore, what would such a probability even mean? Are you postulating a universe-generating machine, randomly spitting out universes, laws and constants?

The short answer: The probabilities that scientists use to test physical theories are *Bayesian* probabilities, and reflect the degree of plausibility of a certain claim, given what we know. We do not need to postulate a real or even imagined ensemble of universes, or a random universe-creating machine. We're trying to think rationally, and probabilities can help us.

The problem of assigning probabilities to a large (or even infinite) number of possibilities can be a thorny one, but it is not unique to fine-tuning. In science, theories compete to explain our evidence. To test theories, we must compare with alternative theories, including different laws, constants and initial conditions. The probabilities that are required to understand the fine-tuning of the universe for life are the probabilities of theory testing in science.

The long answer: The question 'what are probabilities?' is not as straightforward as one might expect. An unbiased coin will be flipped; the probability that the coin will land heads up is ½. What exactly do we mean by that? There are a number of views of probability discussed in the scientific literature, to which we can attach fancy names.[28]

---

[28] We are commenting here on probability as we have seen it in the physical sciences; we are not attempting an exhaustive review of mathematical statistics, epistemology (the study of knowledge) in general or the philosophy of probability in particular. We have, of necessity, borrowed some terms from the philosophy of probability.

Finite frequentism: of all the coin flips that have ever occurred, roughly half have come up heads.

Hypothetical frequentism: if you flipped an unbiased coin an infinite number of times, the fraction of heads would approach a half.

Objective chances: the physical system of your thumb, the coin and the floor will tend to produce a head on half of all flips.

Subjective Bayesianism: before the coin was flipped, my personal belief that the coin would land heads was as strong as my belief that it would land tails.

Objective Bayesianism: given *only* the information 'this fair coin has been flipped', the plausibility of the proposition 'this coin landed heads' is equal to that of 'this coin landed tails'.

Now, all of these statements might be true, and so we are not looking for the *one true interpretation* of probability. For example, if the objective chance of heads is one-half, then this is usually taken to imply that the hypothetical frequency of heads is also one-half. These interpretations differ on whether the probability is assigned to the hypothetical sequence of flips, or to a property of the coin+thumb+floor system.

The two forms of 'Bayesianism' are named for the Reverend Thomas Bayes, an eighteenth-century statistician and Presbyterian minister, and the theorem of probability that bears his name. On his tomb in London is written: 'This vault was restored in 1969 with contributions received from statisticians throughout the world'.

The distinction between the two forms of Bayesianism is somewhat slippery, but goes something like this. The subjective Bayesian aims to quantify the intensity of a particular person's beliefs. They believe that probabilities are loosely constrained by self-consistency, and accept that probabilities reflect non-rational biases and prejudices.

The objective Bayesian, on the other hand, is trying to discover the rules that govern rationality. Edwin Jaynes, whose posthumous textbook *Probability Theory* is fast becoming the bible of probability in the physical sciences, imagined that we are programming a rational robot. We feed the robot certain information ('I just flipped a fair coin'), and then ask it about a statement ('The coin turned up heads'). The robot's task is to assign probabilities in light of *only* the information it is given. Objective Bayesianism aims to extend true-or-false logic to include degrees of plausibility, which quantify the fact that statements such as 'it will rain within an hour' are supported, though not necessarily proved, by other statements such as 'there are dark clouds overhead'. Objective Bayesians believe that probabilities can be tightly (if not uniquely) constrained by principles of reason, and aim to develop methods that overcome personal bias and prejudice.

Philosophers often use the term *credences* for Bayesian probabilities of this 'degree of plausibility' sort, to distinguish them from frequencies and chances. The claim of the objective Bayesian is that, in many situations (and not just in science), we can use general principles of reason to discover the credences of an idealized, rational thinker. If you are rational, your personal credences will be these objective credences.[29]

While debate continues in philosophical circles, the scientific community has largely embraced Bayes. Our search of the NASA Astrophysics Data System[30] revealed over 7,000 physics and astronomy publications with 'Bayesian' in the title. Only about 70 have 'frequentist' in the title, and of those 35 also have the word 'Bayesian' in the title. While frequentist methods are still used, the current dominance of the Bayesian interpretation is obvious. While

---

[29] We will mostly use the term 'probability', as 'credence' is often taken to imply subjective Bayesianism. As mentioned, the terminology of probability is slippery. The divide between *frequencies* (property of a sequence), *chances* (property of a physical system), and *credences* (degrees of certainty or belief) is reasonably standard. However, sometimes all Bayesian probabilities are described as subjective. We use *subjective* in a strong sense: subjective probabilities describe (rather than *prescribe*) the psychological state of a particular individual. For others, *subjective* simply means that probabilities depend on what is known. Calling Bayesian probabilities *objective* is sometimes taken to imply that there exists a unique, correct probability assignment for any hypothesis given any information. Credences are sometimes called epistemic probabilities. Objective credences are sometimes called logical probabilities. Objective chances are sometimes called propensities, though sometimes the two are distinguished – propensities are not reducible to anything more fundamental. The reader interested in learning more is wished the very best of luck.

[30] www.adsabs.harvard.edu

frequentism reigned in statistics just a few decades ago, it is fast becoming the 'f-word' of data analysis in the physical sciences.[31]

A number of features of objective Bayesianism have endeared it to physicists. Most decisively, consider the following claim: given what we know about the Solar System, Einstein's General Relativity is (say) 100 times more likely to be the correct theory of gravity than Newton's theory. Which probability interpretations can *make* such a claim *at all*?

Finite frequentism is out: we're not saying that we keep daily watch on the Solar System and every couple of months it decides to obey Newton for a day. Nor have we observed a population of Solar Systems, and discovered that only about 1 in 100 have a nostalgic fondness for Newton's theory. Hypothetical frequentism fails for the same reason: there is no population of Solar Systems that this probability claims to be about.

The problem with objective chances is that they give us the probability of the *data* given the *theory*. However important and useful these probabilities are, to evaluate a proposed physical theory we want to calculate the probability of the *theory* given the *data*. This represents our situation: we know the observations we have made, and we want to know what theory they support. But theories aren't chancy. Outcomes are chancy.

This inability to even *define* the probability of a theory in terms of frequencies and chances is not denied by proponents of these interpretations. In fact, Ronald Fisher, the patron saint of frequentism, proclaimed: 'we can know nothing of the probability of hypotheses'.[32] This, for Fisher, was a principled refusal to extend talk of probabilities beyond empirical frequencies. To the modern scientist, the most likely reaction is: then what's the point? Why do science at all if we're never allowed to *ask* whether any of our theories are correct, or even plausible?

The subjective Bayesian can state that Einstein's theory is more probable than Newton's, but this merely reports an opinion. Our

---

[31] We recommend the books by Taleb (2010), McGrayne (2012) and Silver (2015) for readers who wish to learn more about modern probability theory and its role in the sciences. Eagle (2011) collects and comments on some important readings in the philosophy of probability.

[32] Quoted in Aldrich (2008).

subjective probabilities could be changed by precise astronomical data, or by waking up on the wrong side of the bed. Scientists have developed methods of overcoming human biases and prejudices, aiming to infer something about how gravity really works, and not just something about how particular human beings respond to images from telescopes. What's the point of science if we can't say something about the Universe, or even something about what we should believe about the Universe?

Given that scientists infer conclusions from data, report our results to the scientific community, and expect our conclusions to be reproducible, we are closer to the *objective* end of the Bayesian interpretation. Objective Bayesianism is coming to dominate statistical practice in physics because of its elegance, its principled foundations, its transparent connection between principle and practice, and its clarity and unity of method. Bayesianism lets us ask the right questions of our data.

This long introduction to probability theory has been essential, as we need to understand how fine-tuning claims rely on the same mathematical apparatus as the rest of physics. Our claim is that fine-tuning claims can be understood in the context of objective Bayesianism, and that the difficulties faced in calculating the relevant probabilities are of the same kind as those faced in analysing any physical theory.[33]

We have presented fine-tuning as a clue to something deeper, something behind the laws of nature as we know them. To see why, we ask: if all we knew was (a) that a certain universe obeyed the laws of nature as we know them, *without* specifying the values of the constants of nature and initial conditions, and (b) mathematical knowledge, what is the probability that that universe would contain life forms? We have seen in previous chapters that this probability seems to be extremely small.

---

[33] This is our response to more technical objections such as the coarse-tuning problem and the normalization problem; see McGrew, McGrew and Vestrup (2003). These kinds of 'what to do with infinity' problems are often encountered in the physical sciences, especially in cosmology, and so these objections cannot succeed against fine-tuning without paralyzing probabilistic reasoning in all of physics.

If that question strikes you as illegitimate, consider its cousins: with the same knowns, (a) and (b), what is the probability that that universe would contain galaxies? Or stars? Or atoms? Or liquid water? These questions are easier than our question about life, of course, but they are the same *kind* of question. They are the questions of theoretical physics, and to test our theories we need to calculate (at least approximately) the associated probabilities.

Of particular concern is the so-called *prior probability* of the constants and initial conditions of a given theory. This is the probability of, for example, a constant having a value in a certain small range, *without* any knowledge about our Universe. That is, we are not inferring the constant from observations, but quantifying the initial fact that the constant is not specified by our theory, and thus without observations we are ignorant of its value.

Specifying prior probabilities is a technical and thorny problem, but one that is not unique to fine-tuning. We need them for *any* calculation of the probability of a theory. In the technical jargon, we treat the constants as *nuisance* parameters, to be averaged over. We even need prior probabilities to infer the most probable value of the constants themselves from our observations. Physicists who want to conclude from their experiments that the electron's mass is, with 95 per cent plausibility, between 510.998 939 9 and 510.998 952 3 keV/c^2 need a prior probability.[34]

---

[34] For those who are interested, here are the mathematical details. Using our preferred physical theory of electrons, you can calculate the *likelihood*: the probability *of* your data (D) *given* the theory and a particular value of the electron mass (m_e). We write this as $p(D \mid m_e B)$. Now, we want to calculate the *posterior* $p(m_e \mid D\,B)\,dm_e$, which is the probability that the electron's mass is between m_e and $m_e + dm_e$, given our data and theory. We calculate this using a continuous version of Bayes theorem,

$$p(m_e \mid DB) = \frac{p(D \mid m_e B)p(m_e \mid B)}{\displaystyle\int_0^\infty p(D \mid m_e B)p(m_e \mid B)\,dm_e}$$

We cannot calculate the posterior at all without *some* estimate of the prior probability $p(m_e \mid B)\,dm_e$, which is the probability that the electron's mass is between m_e and $m_e + dm_e$, given the theory *alone*, without taking into account any data.

If our data are very good, then our conclusions won't depend much on the prior probability. Similarly in fine-tuning cases: the speed and severity with which disaster strikes as one tiptoes through parameter space show that the probability of a life-permitting universe, given the laws but not the constants, will be very small for any honest (and non-fine-tuned!) prior probability.

Recall that this probability represents our ignorance of the parameter, not any property of a random universe factory. The fact that we don't know what sets the constant is not a deal-breaker; it is precisely what we're investigating.

We have already noted that our search for life in hypothetical universes has predominantly (though not exclusively) taken us to worlds that are similar to ours. For example, most fine-tuning cases consider universes with the same laws but different free parameters. Does this bias our search? Yes ... *in favour of finding life*. As an illustration, it's a good idea to look for mushrooms near other mushrooms. If the conditions are just right for mushrooms around a particular tree in the forest, then nearby trees are more likely than some randomly chosen tree to also be right for mushrooms. Similarly, looking at universes that are similar to ours prejudices our search *in favour* of finding life-permitting universes. It is all the more surprising, then, that life-permitting universes are extremely rare even in this biased sample!

Let's summarize. We haven't made a watertight argument here, much less performed a calculation. We certainly don't have a perfect, patented prior probability producer.[35] Rather, think of this as a *threat*: the probabilities of fine-tuning and the probabilities of theory testing in physics are inseparable. Bayesian theory testing is nice. Very nice indeed. It would be a real shame if something were to ... *happen* to it.

---

[35] With apologies to the Anglo-Australian Anti-Alliteration Association.

A Final Few Reactions

A few reactions to fine-tuning remain. They come up the most often, and are where most of the debate about fine-tuning is today.

The first is that the constants of nature will be explained by deeper physics. When we really understand the laws of nature, we'll know why the Universe could not have been different, why these constants couldn't *really* have taken on life-prohibiting values.

The second is the multiverse: there is a vast, varied collection of universes out there, in which the right set of constants is bound to turn up somewhere. We must observe a life-permitting universe because those are where all the observers are!

Finally, a designer. The Universe has its properties because it achieves the goals of a universe creator. We're in a universe that permits the existence of intelligent life forms because that seemed like a good idea to someone; in particular, someone with the knowledge required to suitably design the laws of nature and the power to bring such a universe into being.

These reactions to fine-tuning require more than a brief answer. They will be the focus of our final chapter.

8   A Conversation Continued

You should congratulate yourself on reaching the final chapter of our book. It has been a long journey, from the physics of the very small to the workings of the entire cosmos, with bits of everything in between.

The message: messing with the make-up of the Universe can have a disastrous effect on the emergence of complex life like you and me, and especially the physical conditions that underlie life, such as useable energy and organic chemistry. Our conclusion is that the fundamental properties of the Universe appear to be fine-tuned for life. We need a cosmos that expands not too fast and not too slow, that forms structure, with a mix of stable elements that can form stars, planets and cells, with the right mix of forces for stars to burn for billions of years, with plenty of carbon and oxygen, with a low-entropy past and free energy into the future, with a life-supporting number of dimensions, and even with mathematically elegant and discoverable laws. Such a cosmos is a rarity among our Universe's cousins and distant relatives.

This should be quite startling! Sitting here on the surface of this small rocky planet, with the Sun beaming down and rain clouds on the horizon, or in the hustle and bustle of a busy street, or walking through the silence of a snow-covered landscape, our existence seems so, well, natural. However, we appear to be the result of a cosmic gamble that makes winning EuroMillions look like a dead certainty – a rather unsettling thought!

Well . . . now what?

We would like to know: why is the Universe like this? We cleared away some distractions in the previous chapter. But still you may be wondering if this is a question we can even start to answer. The ideas inspired by the fine-tuning of the Universe for life range from realistic science to informed guesswork to unfettered speculation. It is often hard to tell which is which.

In this final chapter, we return to our two chatting cosmologists from the start of this book. At the risk of spoiling the suspense, we haven't got all the answers. We'll *start* the conversation; find some good friends and some good beer and carry on when we're done.

THE CONVERSATION CONTINUES

Narrator: The day has worn on, and the Sun is setting. The two cosmologists are still discussing the fine-tuning in the Universe for life. They've spent the day chatting about quarks and electrons, dark matter and spacetime, and have come to agree that the Universe could have been significantly different. It could quite easily have been significantly dead. They are starting to wonder if we can explain *why* our Universe appears so finely tuned for life like us.

Geraint: We've discussed a lot of things today. The best theories of how the Universe works have a few loose dials. And, spurred on by curiosity, or mischievousness, we've asked: what would happen if we gave those dials a spin? What weird and wonderful worlds might result?

Playing with the laws of physics, it turns out, can be catastrophic for life. Often, the catastrophe is boredom. The periodic table disappears, and all the astonishing beauty and utility of chemistry desert us. The galaxies, stars and planets that host and energize life are replaced by lethal black holes or just a thin hydrogen soup, lonely protons drifting through empty space, and a bath of tepid radiation. These are very dull places indeed, and not the kind of place that you'd expect to encounter complex, thinking beings like us.

Luke: In fact, it is difficult to imagine how life of any kind could arise. For example, a careless bump of the dials that control the masses of the key fundamental particles, the up and down quarks and the electrons, results in matter that is too simple to make anything.

Geraint: Not only that, but it seems that we were lucky to find ourselves in a Universe that was born in a very low entropy state, with plenty of free energy to power the various processes necessary for life to arise. The thought of our baby Universe being filled with nothing but black holes, with very little free energy, leaves me cold. In fact, I shudder at the thought of all of these potential universes appearing to be completely sterile. I guess we should feel pretty fortunate to be in a universe that can sustain us!

Luke: And don't forget the many universes that recollapse rapidly after their birth in a Big Bang. While such universes may have had the ingredients for life, their short existence means that there is no time for stars, planets, complex molecules and life before the end comes, before being crushed out of existence. And still others would have expanded too quickly, leaving matter too dilute to form stars and galaxies, or indeed any structure whatsoever.

Geraint: A pretty grim picture. And yet, here we are. This Universe appears to have the right properties, the right recipes, the right laws of physics, to allow life to flourish on at least this one planet, and possibly on the many of the other planets that astronomers are discovering. It does make you wonder, however, why our Universe is the way it is.

Luke: And at this point we have to be careful. We're moving away from science as we usually practice and picture it. Speculation and guesswork await. So let's try to be systematic about exploring the options.

THE LADY GAGA DEFENCE

Geraint: One solution that immediately springs to mind is that the Universe just is and that's all there is to it. Like Lady Gaga, it was just 'born this way'. It simply is what it is.

Luke: But I've still got questions! Science has come an awfully long way with its 'keep looking for answers' attitude. Why is the electron mass what it is? Why three generations of quarks? Why is the Universe as lumpy as it is? What caused inflation, if it happened?

Geraint: These are, of course, excellent questions. But what's wrong with just simply saying, 'They are just so!' After all, our explanations presumably have to stop somewhere.

Luke: It sounds like a parent ending a child's wearying series of 'Why?' questions with 'It just is, OK!' This isn't an answer at all. It's fine to say, 'I don't know'. It's something else entirely to say that the question has no answer. Not just that I don't currently know the answer or that no one knows or even that no one could know. 'It just is' says that there is no true fact that answers the question. You could write down every true fact on a great big piece of paper and nowhere would you find 'there are three generations of quarks because ...'

Now, short of knowing everything, how would we know that a question has no answer? When do we call off the search? Jim Holt's entertaining book *Why Does The World Exist?* gives the following provisional guide to inquiry: 'Always look for an explanation unless you find yourself in a situation where further explanation is impossible' (2012, p. 237).

Geraint: That sounds reasonable, and I guess it shows what's wrong with 'they are just so'. These questions about the Universe don't seem impossible to answer. In fact, science has answered these kinds of questions before, at least in part. We have options, and until they fail we should keep looking for an answer.

So, ultimately I am not very satisfied with the Lady Gaga defence. It feels like we are ignoring the problem, not even trying to answer it. Blissful ignorance, but without the bliss. But just what should we be thinking about?

MORE PHYSICS, PLEASE

Luke: Remember what these dials are. The best theories we have about the basic stuff of the Universe, how it behaves and how it is arranged, need a few numbers or *parameters*. The laws need to be told how much an electron weighs, how strong the forces of nature are, and how much matter is in the Universe. With these numbers, our equations predict a mountain of data about how the Universe behaves: how electrons move in a wire, how light warms your skin, how planets orbit a star and how space expands. These numbers are vitally important to our understanding of the workings of the Universe, but they cannot be calculated, only measured. We don't know why they are what they are.

These *constants of nature* are of intense interest to physicists, because they offer a direct route to *deeper laws*.

How do we find clues to deeper laws of nature? Maybe new experiments, but these are expensive and risky. We could show that our existing theories don't account for some data, but this is fraught with ifs and buts; are we solving the equations correctly, or do we fully understand the details of our experiments?

But if you've got a new theory that can simply and naturally predict the value of a fundamental constant, then you're on to a winner. It predicts something that *by definition* the old theory cannot handle.

Geraint: So, do we just need more physics? In looking more deeply at the workings of the Universe, perhaps we will learn that nature's dials are not really free to be twisted, but are stuck at the value they have.

Luke: Einstein stated this idea – a dream, really – with his usual clarity[1]:

> I would like to state a theorem which at present can not be based upon anything more than upon a faith in the simplicity, i.e. intelligibility, of nature: there are no arbitrary constants ... that is to say, nature is so constituted that it is possible logically to lay down such strongly determined laws that within these laws only rationally completely determined constants occur (not constants, therefore, whose numerical value could be changed without destroying the theory).

---

[1] In Schilpp (1969), p.63. Yes, Luke really does use footnotes in conversations.

Einstein was a smart man! In the laws of nature, as we currently know them, we can spin the dials without destroying the theory. The theory still makes predictions. Mathematically, everything still works. Einstein, along with every other physicist, would prefer a theory with fewer free parameters. We're always on the lookout for simplicity, and a theory with zero free parameters seems very simple indeed.

Geraint: OK, I get it. The hope is that the progress of science will eventually do away with these constants, and so do away with the knobs and dials with which we've been fiddling. We will find that properties like the electron mass are fundamental and immutable, laid down by the theory itself. The Universe will emerge, fully formed, from some 'theory of everything'.

If this is the case, we will realize that asking the electron to be heavier is like asking two and two to equal five. Maybe someday we'll understand why things couldn't have been otherwise. Is this game over?

Luke: There's a lot to say for Einstein's idea. It's what physicists daydream about – to sit down with pen and paper, write down some simple equations and from them understand how our Universe is put together. Even if we couldn't solve the equation perfectly, nature would be to us, in words of the brilliant Austrian physicist Ludwig Boltzmann, a difficult problem, but not a mystery.

However, the dream is not so simple. For a start, the last few centuries of physics have not seen a consistent trend towards fewer fundamental constants. Sometimes, a new theory unifies some constants. One of the most famous of these unifications was the work by James Clerk Maxwell in the late 1800s, showing that electricity and magnetism are actually two sides of the same coin. He showed that the constants of electricity and magnetism are related to the speed of light. This is great physics, and the stuff of scientific legend. But at other times, new discoveries have required new fundamental constants.

Our best hope for a theory with no free constants has turned out to be less than encouraging. *String theory* is believed by some to be the future of fundamental physics, uniting gravity and quantum mechanics into a single framework. It was hoped that string theory would be able to calculate the values of any physical quantity of interest, including the fundamental constants.

However, this hope was dashed when it turned out that string theory is not as 'strongly determined' as Einstein dreamed. What are constants of nature to us – numbers in the equation – are to string theory initial conditions – numbers in a *solution* to the equation. But they are free parameters nonetheless, and a far cry from Einstein's 'completely determined constants'.

Geraint: So Einstein's dream is still just that, a dream. And there is a real prospect that it might remain so. Thinking about it, it seems worse than you've made it out to be. Even if Einstein's dream were true, it *wouldn't* show that the Universe couldn't have been otherwise. It only shows that the constants are determined within that particular deeper theory. We would still want to know why the Universe obeyed that deeper theory.

Even if there are no alternatives *within* the deeper theory, there are alternatives *to* the deeper theory.

Luke: I agree! If Einstein's dream were true, instead of writing books about the fine-tuning of the constants of nature, we'd write books about the fine-tuning of the laws of nature themselves. Instead of alternative values of constants within a theory, we'd discuss alternative theories. But all the same fine-tuning problems would arise. Bernard Carr and Martin Rees made this point succinctly in 1979:

> Even if all apparently anthropic coincidences could be explained in this way [by some presently unformulated unified physical theory], it would still be remarkable that the relationships dictated by physical theory happened also to be those propitious for life.

This raises an even more fundamental issue. Go to a whiteboard and write down any equation you like. All sorts of consequences may flow from that equation. If you scribbled down Newton's law of gravity, one consequence would be the orderly orbits of the planets around the Sun, something that we see in the Universe. But other potential outcomes, such as planets moving 100 times faster than the speed of light, are something that we don't see. The fact that something is described by an equation doesn't mean that it is out there in the Universe. More generally, no equation, however beautiful, simple or self-contained, can command a universe's obedience.

This is a sobering conclusion. You can't start with just an equation and conclude that electrons exist. The question of what *must* exist and what properties it *must* have is not one that can be answered by the usual methods of science. If you want to say that life-prohibiting universes are impossible, you're going to have to give the kind of reasons that science cannot give.

Geraint: I think I understand what you are saying, though it's hard to think about the Universe without imagining a process that brings it into being and imprints its properties.

Maybe the problem is that we are treating this Universe as unique, and maybe fine-tuning is not such a problem if we just step back a bit. What if the process that gave birth to the Universe churned out many more?

Luke: I've heard this one before, but it can sound like crazy talk. Are you going to play the *multiverse* card?

Geraint: Precisely! Let me explain.

THE MULTIVERSE

Geraint: So, while I don't think that physics will uncover a unique set of properties underlying the Universe, in that this is the one-and-only-way a universe could possibly be, I do think it provides a possible solution to fine-tuning.

Luke: I'll play the 'curious bystander'. How so, my learned friend?

Geraint: Well, remember that we think that the Universe underwent a rapid burst of expansion soon after it was born, a period known as *inflation*. Inflation irons out some of the problems of the standard Big Bang cosmology, providing an explanation of why it appears to be spatially flat, as well as fixing that pesky horizon problem.

While inflation is an impressive idea, there are a number of competing mechanisms that try to explain how the Universe could have undergone such rapid expansion. The details vary – how inflation started, how it proceeded, how it ended, and, of course, what caused it – but in many of these, different parts of space undergo differing amounts of inflation. In our local patch, expansion could be brief, turning on and off in an instant. But in other patches, inflation continues. The Universe will never run out of inflating space.

The result is a vast number of other patches, so enormously distant from us that we can think of them as separate universes. In some, inflation was briefer than in ours, in others it was much more drawn out, and in some it will never stop.

If the 'constants' of nature are shifted and shuffled in extremely high-energy conditions, then the early Universe will provide each inflating region with a different set of constants. Observers in a patch – *if there are observers* – would see different particles, forces, and more.

This grand ensemble of other universes is known as the *multiverse*. At this point, we must be careful with our language. A dictionary definition of 'universe' means every physical thing, all space and all time. However, in the context of the multiverse, the picture becomes murky, as we can now refer to individual patches within the multiverse, patches with their own laws of physics, as individual universes. And as with all science at the edge, there are still a lot of unknowns.

Part of the problem is that there is no single agreed definition of what a multiverse is. To some, the multiverse is nothing more than differing patches within our Universe, with the laws of physics imprinted by inflation. To others, the multiverse is the result of quantum mechanical decision-making, the much-discussed Many Worlds idea. And some even think that any mathematical structure counts as a real 'universe', so that every possible universe is, in some sense, out there. I am sure that there are many other possible multiverse ideas that we haven't even thought of yet.

The multiverse makes some people uncomfortable. It seems far too speculative for science, and its picture of reality can be unsettling. Our Universe would be just one amongst many. In some universes, black holes rule, or the electron has no mass, or planets spiral into their stars. As we've seen again and again, the vast majority of these universes will be stone-cold dead, sterile, barren places where life has no chance of arising and thriving.

Luke: Being cosmologists, why don't we confine ourselves to multiverse ideas that are at least motivated by our observations of this Universe and the need for rapid expansion soon after it was born? This inflation was proposed to solve various problems in *cosmology*. It was supposed to make universes that were approximately flat and smooth. But that's not enough to solve fine-tuning; we need to spin both the cosmological and particle physics dials. Why think that inflation will cause such a kerfuffle that the laws and constants of nature will be scrambled?

Geraint: Actually, it's better to think of inflation as smoothing out the kerfuffle. The laws of nature as we know them get more *symmetric* at higher energy. By this, I mean that the various forces, the strong, the weak and electromagnetic, start to look very similar when we consider very energetic collisions of particles. Extrapolating, in the super-hot melee of the early Universe, we think that the various forces were actually one and the same.

But as the Universe expanded and cooled, it reached temperatures where the laws of nature could not maintain this perfect symmetry. The symmetry is *broken*, and it is only when this happens that the 'constants' of nature (as we know them) take on a particular value.

Here's an illustration. There is nothing about cars and driving that favours driving on the left versus the right.[2] There is *mirror* symmetry. The important thing is that when this symmetry is *broken*, when we choose which side of the road we will drive on, we all do it *together*.

Luke: A joke ... to illustrate. Keith walks into an Australian outback pub.

KEITH: I wouldn't go to America if you paid me!

BARMAN: Why not, Keith?

KEITH: Well, they drive on the other side of the road over there.

BARMAN: What's wrong with that?

KEITH: I tried it the other night and it's bloody dangerous!

[Crickets chirping in the background.]

Geraint: Moving on. Think of a box of magnets. If you shake the box vigorously, the magnets will bounce around, snapping together and breaking apart. Most of the time, they will be flying and spinning freely. There will be no net magnetic field from the box, as the magnets show no preference for any particular direction.

Now, gradually slow things down. Collisions with the sides of the box and other magnets are less likely to jolt the magnets apart, and so groups develop as the magnets snap into a tight formation. Like picking a side of the road, the magnets will tend to pick *some* direction and align accordingly. Any rebellious magnets will be pushed into line by the combined force of all the others. The broken symmetry is enforced.

Suppose opposite ends of the box pick different directions. They will begin to enforce their choice on their surroundings, creating larger formations with a yet stronger collective magnetic field. When these growing formations meet in the middle, a stalemate is reached, with neither able to make the other budge. The box will fragment into domains, each having broken the symmetry differently.

---

[2] The predominance of right-handers is an exception, but even this isn't particularly decisive.

The same kind of process explains the formation of snowflakes. In warmer air, water exists as vapour and droplets, all identical. As the temperature drops, individual snowflakes begin to form as the water changes *phase* from gas to solid. While the physical process is the same for each of them, tiny differences in the starting pattern, such as a piece of dust here or a shard of ice there, means that each snowflake is unique.

Luke: These analogies are fine, but what in the early Universe corresponds to the magnetic field or the water vapour in these examples? What changes phase?

Geraint: Inflation itself is a good example. The *inflaton* – whatever form of energy caused the very early Universe to expand so rapidly – undergoes this abrupt kind of change when inflation ends. Different parts of the Universe end their inflating phase at different times, creating the patchwork of domains I mentioned before.

To apply these ideas to the fundamental constants, we're going to have to give up the idea that they are *constant*. We must suppose that they are really something else. Our equations are merely approximations to some deeper, grander equations in which the 'constants' are dynamical entities such as fields.

When the Universe breaks a symmetry of these fields, it can splinter into domains. Each has a somewhat random mix of field values, as unique as a snowflake. To the inhabitants of each domain – if there are any – these frozen-in fields are observed as the constants of nature.

The domains will be of varying size, but are likely to be miniscule compared to the Universe that we observe. But that's where inflation comes in: some of these domains are stretched to become Universe-sized and larger. And if inflation keeps bubbling away, there will be a vast multiverse of such regions, each with different laws. Inflation ends up being like the labourers in the 1920s movie *Metropolis*, furiously spinning the dials, and each time they stop, a new universe is made, with its newly minted, and quite distinct, laws of nature!

Luke: Well, congratulations if you got that reference.[3]

Geraint: Who are you talking to?

Luke: The crickets.

Let's get into specifics. Does anyone have any idea how to change our constants into physical, dynamical things?

Geraint: Our best candidate, in the view of many, is *string theory*, as mentioned before. Every piece of matter and light, if we could look closely enough, is a vibrating 11-dimensional string. The appeal of this idea – behind all the maths[4] – is that it needs only one kind of thing. Instead of postulating about 20 kinds of particles, the different patterns of the wobbling string make it behave like different particles.

String theorists hoped that, starting with the right equations, they would be able to derive all the properties of our Universe. The 'constants' of our equations would be predicted by string theory. Unfortunately, it wasn't to be. While there is hope that the equations of string theory will be unique (and there's a warning bell: we don't even know the full equations of the theory, only approximations), the solutions to the theory are a vast, untamed horde. What in current physics are free parameters in our equations become free parameters in the *solution* to the equations of string theory. They have changed status – from constants to initial conditions – but remain as free and unpredictable as ever.

Luke: What do these solutions look like?

Geraint: At some level, they're going to have to look like our Universe, which you may have noticed is not 11-dimensional. Our Universe has three dimensions of space and one of time (physicists write this as *3+1 dimensions*). String theory is free from internal contradictions in 11 dimensions, and so to create our familiar 3+1-dimensional world, the extra dimensions must be *compactified*. I'll explain.

---

[3] For those who have not seen this classic movie, the scenes are reconstructed in the equally classic music video for *Radio Ga Ga* by Queen.

[4] Or *math*, if you prefer. Anyway, back to the *physic*.

In 3+1 dimensions, we need three numbers to tell us *where* something happened, and one number to tell us *when*. For example, the party is on 7th Avenue, number 24, 5th floor at 7:30 p.m. A dimension can be *infinite* or *compact*. We think of space as infinite – there are an infinite number of metres out there for you to travel. The surface of the Earth, on the other hand, is compact – there are only a finite number of metres you can travel in a straight line before you get back to where you started. Compact dimensions can be large or small, depending on the space available.

A century ago, physicists realised that if there were small, compact extra dimensions to space and time, then they would not necessarily be observed. Like single atoms, they would be too small to see. Then why bother, you might ask.

Indeed, extra dimensions would be science fiction, but for an oddity noted in 1921 by the German physicist Theodor Kaluza. He wrote down Einstein's equation of General Relativity for a universe with an extra, compact dimension. He found that the extra dimension generated an extra equation, which – astonishingly – was the equation for electromagnetism! Oskar Klein showed in 1926 that applying quantum mechanics to this extra dimension would explain why electric charge is quantized, that is, why the charges of elementary particles come in whole number units of some elementary charge. Maybe gravity and electromagnetism aren't separate forces, but are unified in an extra dimension. Could the other forces and particles be included with a few more dimensions?

This *Kaluza–Klein theory* is the start of the road that leads, a century later, to 11-dimensional strings.[5] We still don't know if the idea is correct, but strings have certainly captured plenty of attention from physicists.

Luke: OK. We were talking about *compactification*: we've got an 11-dimensional theory, and we need to hide all but 3+1 (3 space + 1 time) of those dimensions.

---

[5] We cannot tell the whole story here. Brian Greene's *The Elegant Universe* (1999) is a splendid introduction to string theory.

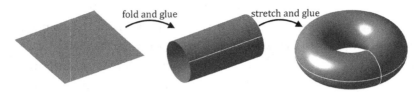

FIGURE 47 How to make a torus, which is a particularly simple example
of a compact shape. Take a stretchy rectangle of rubber. Fold the long
edges together and glue. Then stretch it around and glue the ends together.

Geraint: Right. So, attached to each point in our familiar space
and time is a higher dimensional shape, in which the extra dimensions
are folded up. We cannot see these dimensions, but the strings can.
And they affect how the strings wobble, which determines the kinds
of particles that we see in our experiments.

So, the 'constants' of nature are all written in the compactification. The shape of the extra dimensions determines the physics of the
particles we see around us.

String theorists began by considering simple shapes. Think of a
stretchy rectangle of rubber. Fold the long edges together and glue.
Then stretch it around and glue the ends together. The result is known
as a torus (see Figure 47). This can be done in higher dimensions,
creating a particularly simple example of a compact shape.

So, what kind of particles do strings look like when they wobble
around a torus? Remember that particles can be right or left handed,
depending on how they spin. Our universe contains particles whose
right-handed versions look different from their left-handed versions. It
turns out that this isn't possible for strings on a torus, so we need to
find another shape.

Thus began a massive effort by a community of physicists and
mathematicians to try to find suitable higher-dimensional shapes to
put into string theory. The left–right-handed problem (*chirality*, in the
jargon) can be solved using a special shape known as a 'Calabi–Yau
manifold'. Such a shape is characterized by various parameters, such
as the sizes of the dimensions, the number of holes in the shape, etc.

Some of these shapes predict particles that we do not observe, known as *massless scalar particles*. Some are perfectly *supersymmetric*, which our Universe is not.[6] Some are simply unstable.

Interestingly, if we avoid those undesirable shapes, the remaining possibilities seem to be finite in number. That number, according to the most popular estimate, is around 10^{500} – that's 1 followed by 500 zeros. This vast set of possibilities has been called the *string theory landscape* by Leonard Susskind of Stanford University.

Luke: But the landscape is not a multiverse. It's just a set of possibilities.

Geraint: Correct. A large set of possibilities can't *of itself* solve the fine-tuning problem; in fact, it's part of the problem. As an illustration, the large number of *possible* lottery tickets is precisely what makes winning unlikely. To have a better chance of winning the lottery, we need a large number of different *actual* tickets to be purchased.

It's worth quoting Susskind himself on this point, as it is often misunderstood:

> The two concepts – Landscape and megaverse [i.e. multiverse] – should not be confused. The Landscape is not a real place. Think of it as a list of all the possible designs of hypothetical universes. Each valley represents one such design. . . . The megaverse, by contrast, is quite real. The pocket universes that fill it are actual existing places, not hypothetical possibilities.
>
> (2005, p. 381)

So, we need something like inflation to actually populate the landscape. In an inflating universe, each of these possibilities may be realized, probably many times.

Luke: So, getting to the crux, where is life in this hullabaloo?

Geraint: Here's where the anthropic principle is actually useful. We find ourselves in a universe that is fine-tuned for life because the

---

[6] Recall from Chapter 3 that 'supersymmetry' implies that each particle has a cousin with the same properties except for spin. We don't see this.

properties of this universe allowed life to arise. We would not expect to find ourselves in a universe that existed for only a matter of seconds, or one in which there were no protons, neutrons or even chemistry.

Luke: Given what we know about fine-tuning, how many universes do we need to create to beat the odds and ensure that a life-permitting universe is amongst the many produced?

Geraint: That's a hard question, as we are still only really coming to grips with the physics of fine-tuning. As we have seen, there seems to be an awful lot of ways of making a dead universe, so a multiverse had better make a lot of universes if it wants to produce one that allows life. In principle, we could answer this using the string landscape – just work through all the possibilities and note what fraction might allow for life. But this is impossible in practice: there are far too many and the calculations are too difficult.

Luke: William of Occam famously stated, 'plurality must never be posited without necessity'. Is positing an awfully large, though not necessarily infinite, number of universes a violation of *Occam's razor*?

Geraint: We have to ask Mr Occam, a plurality of *what*? Things? That doesn't seem right. According to atomic theory, this room contains of a couple of trillion trillion atoms. But we don't penalize the theory because of the huge number of atoms.

It is *independent assumptions* that shouldn't be multiplied beyond necessity, I'd say. Every theory begins with a set of basic statements, from which its predictions follow. We prefer fewer assumptions, because it tends to produce simpler theories with fewer basic kinds of things and fewer laws that they obey.

So, the multiverse will avoid the sharp edge of Occam's razor if it needs only a few assumptions. A multiverse theory that orders one universe at a time, like a hungry and haphazard diner at a restaurant, is terribly complicated. But if we find a relatively simple universe *generator*, then the multiverse's vast and varied horde would come relatively simply. If we could borrow a few of the generator's pieces from known physics, then the number of novel postulates is even smaller still. Modern multiverses are like this; cosmologists who entertain thoughts of the multiverse are really looking for a universe generator.

Luke: Let's think about these other universes, then. Firstly, where are they?

Geraint: Well, inflation has separated us from them by immense distances, much larger than the size of the observable Universe. And due to the continual expansion, the light from these other universes will never reach us. They are unobservable.

Luke: Well, that's a blessing and a curse. It explains why our Universe appears to have the same properties everywhere we can see. The physical properties like density, physical constants like the mass of the electron, and physical laws of a typical universe would be the same over very large regions, thanks to inflation.

But if the history of science teaches us anything, it's that testing our theories with experiment and observation is not an optional extra. We aren't very good at guessing what the Universe is like. We have to go and look. Does the multiverse make any predictions that we can test?

Geraint: We have to slow down a little here, because while we have some good ideas about inflation, a few crucial details are missing. We don't know what the inflaton is. Also, we are dogged by the same problems we always find when looking at the Big Bang, namely that General Relativity and quantum mechanics won't work together.

But once we've overcome that little hurdle, we can predict what kind of universes inflation creates. We will still have parameters to play with, but these will be the parameters that govern inflation, and spinning these will result in a different collection of universes in the multiverse. And then we can start talking about predictions!

So imagine one setting on the inflationary dial. This particular setting would give its own peculiar multiverse. But if it is an acceptable description of nature, as we know it, then it certainly has to produce the Universe we currently inhabit! If your theory can't do that, or if it seems very unlikely to do that, then this particular dial setting is destined for the scrapheap of unsuccessful theories.

Luke: That's too weak. Given that we've started from the laws of nature as we know them, and then added a few extensions and appendages to make it inflate, it shouldn't be too surprising if our Universe is somewhere in the mix.

Geraint: Not necessarily. If a multiverse theory predicts that a number of 'constants' vary over a wide range, but only actually generates 17 universes, then the probability that any are life-permitting is very small.

Luke: OK, but there must be more to making predictions with a multiverse. If some multiverse created plenty of life forms, but none of them observed – as we do – three generations of fundamental particles or black holes at the centres of their galaxies, then that multiverse would be ruled out.

Geraint: Right. Making life *somewhere* is what the multiverse does best – just keep cranking out universes until you get lucky. If your aim is terrible, and you're aiming at a very small bullseye, then you'd better throw a lot of darts.

But when it comes to predicting observations, the multiverse doesn't get a free ride. Within the bounds of the anthropic principle – that observers must inhabit an observer-permitting universe – multiverse theories will differ on which kind of observers and which observations are most likely. For example: how old is the universe of a typical observer? How lumpy? How large a cosmological constant? How many particle species? How many neutrinos?

Here's a simple example. Suppose I bought a *blue* shirt from one of two store catalogues, but I can't remember which one. I'm very fussy about size, but if both stores have a large enough range then it isn't surprising that they have a shirt that fits me. And, given that I bought a shirt, *of course* I'm going to buy a shirt in my size.

On the other hand, I'm not fussy at all about style and colour and such, and will choose at random from the shirts that fit. If 30 per cent of the shirts that fit me in Catalogue One are blue, but only 3 per cent of the shirts that fit me in Catalogue Two are blue, then it's more likely that I bought my blue shirt from Catalogue One.

FIGURE 48 On Infinite Street are countless houses, alternately painted black and white. You wake up in one of the houses. Is the probability that you are in a black house one-half, or is it 'infinity over infinity', which isn't really an answer?

Now, just think of shirts as universes. Any multiverse theory, to be worthy of the name, should give us a catalogue of universes that would exist. Given that we are observing the Universe, then *of course* we bought one in the 'observer-permitting' section. But what about the style or colour? If 30 per cent of observer-permitting universes in Catalogue One have, say, three particle generations compared to 3 per cent of those in Catalogue Two – and we are otherwise ambivalent between the two – then we should prefer Catalogue One.

Luke: Some of these multiverse theories postulate an infinite number of other universes. What's 30 per cent of infinity?

Geraint: When physicists propose a theory, whether about electrons, thunderstorms, galaxies or universes, they must be able to calculate the probabilities of possible observations. For example, if you think you know how the Solar System works, then use today's data from the telescope to predict what the night sky will look like tomorrow.

So, if we're going to take it seriously as a physical theory, a proposed multiverse theory must justify its probabilities. Now, infinities are a problem, because we can't just count up the various kinds of universe. We'd just get 'infinity over infinity', which isn't a number. Cosmologists call this *the measure problem*.

However, sometimes these infinities don't seem so hopeless. Suppose that you wake up in a house on *Infinite Street* (Figure 48), where the outsides of the houses are alternately painted black and white, and the street is infinitely long.[7]

If this were *Finite Road*, with a limited number of houses, then the probability that you are in a black house is one-half. This is true no matter how long Finite Road is, so it seems like we should say that the probability of being in a black house on Infinite Street is also one-half. Think about it like this: would you pay a dollar to guess the colour of your house, winning $100 if you guess correctly? That seems like a good bet.

But if you try to take the ratio of the number of black houses to total houses on *Infinite Street*, you get a probability of infinity over infinity. You can't do anything with that. If you don't know your chances of winning the $100, then you don't know if you should take the bet.

These infinities are trickier than they look. I can rearrange the houses on *Infinite Street* to make, say, every 1,000th house black. I haven't lost any houses; they're all somewhere. But the pattern is now ... 999 white, 1 black, 999 white, 1 black ... *everywhere* on *Infinite Street*. Now the $100 prize doesn't look like such a good deal.

The problem is this: we can't assign probabilities on *Infinite Street* in a way that doesn't mind if we then shuffle the houses. So, we have two options. We can demand the right to shuffle, and so give up on *Infinite Street*, refusing to say anything about what is or isn't likely. Alternatively, we can give up the right to shuffle, and look for a way to treat *Infinite Street* as if it were a really long *Finite Road*.

Applying this to an infinite multiverse, the enterprising cosmologist seems to have two options: either provide a convincing way to measure those troublesome infinities, or else the theory fails to provide predictions and so should be discarded. Either option is interesting, and cosmologists are still debating.

---

[7] This example comes from a talk by Cian Dorr and Frank Arntzenius called 'Self-locating beliefs in infinite worlds', delivered at the Philosophy of Cosmology Conference, Tenerife 2014 (youtu.be/OtxZ_wLb_84).

Luke: OK. Enough with the illustrations and analogies. Let's see an example.

Geraint: In the late 1980s, the Nobel Prize-winning physicist Steven Weinberg was pondering the cosmological constant, and in particular the contribution from vacuum energy. Remember that this is a decade before observations of distant supernovae provided good evidence for the cosmological constant.

Weinberg was wondering: what if the Universe had a cosmological constant? As we saw in Chapter 5, a small cosmological constant wouldn't significantly influence the Universe and its inhabitants. But if the strength of this universe-accelerating force were larger, it would prevent the formation of stars and galaxies.

So, in his mind, Weinberg pictured the multiverse, with its patches with differing values of the cosmological constant. In most places, the cosmological constant would make space expand too fast or recollapse too quickly for galaxies to form. In a narrow range, galaxies can flourish, and it is here that we would expect to find our Universe.

A decade later, cosmologists discovered that not only is there a cosmological constant, but it's about the same size as Weinberg predicted! A successful cosmological prediction based upon the idea of the multiverse!

Other scientists have tried to produce similar predictions for other constants, with varying degrees of success. This is a hard problem, but one that physicists and cosmologists can attack.

Luke: So, we can give some multiverses a pass – provisionally. Can we fail them?

Geraint: Absolutely. In fact, many have!

An inflationary multiverse itself flirts with failure in the following way. Consider two identical patches of the Universe, called A and B, that are inflating, doubling and redoubling their size in an amazingly short period of time. They are in a contest to see who will form the most life by some later time, say, a billion years later. But life can't form while the region is still inflating – there's no matter! They'll have to stop inflating, produce matter and have that matter collapse into galaxies and stars and planets and so on.

So, what should their strategy be? Region A decides to stop inflating now, to give it a bit more time before the deadline to create life. Region B decides to inflate for just a bit longer, hoping that the delayed start will be compensated by more space, more matter, more stars and planets, and so hopefully more life.

Who wins? There are two relevant timescales. There's the time needed for a universe to create galaxies, stars and life, which is at least millions – if not billions – of years. There is also the time for inflation to double the size of the universe, which is roughly 10^{-38} seconds. So, if B waits one second longer than A, it will have one second less to form life, which will make almost no difference, and a patch of universe that has *doubled* in size 10^{38} times more than A. It's not just 10^{38} times larger; it's 2 to the power of 10^{38} times larger. That's a number with more than a trillion trillion trillion digits.

So, B wins by a preposterous margin. At any particular time, one-second-younger observers outnumber us by squillions to one. Most observers are freakishly early risers in a young universe, popping up on the first planet available. The inflationary multiverse seems to predict that we are much more likely to be younger rather than older. But, old we are, and so there goes the theory.

Well, not quite. In fact, inflation *survives*. The problem with the argument above is that it considered the entire universe *at a particular time*. This will not do, for two reasons. Firstly, relativity prevents any attempt to define a universal 'now'. Secondly, it seems like we should be considering the Universe as a whole both in space and time. We should survey not just observers 'now', but all observers past and future. Freakishly early risers are in universes that will presumably go on to make observers like us in the usual galaxy-star-planet-evolution style way, and make more such observers in the usual way. Looked at in this way, the oldies might win after all.

This is another aspect of the *measure problem*: how do we survey the multiverse? More precisely, what groups of observers do we compare in order to make predictions? This is why I said that inflation *survives*, rather than *succeeds*.

Luke: Are there any clear cases of failure?

Geraint: Absolutely! Epic failures. Now, I warn you, this is going to get a little strange.

Arguably, Ludwig Boltzmann, who we met a few moments ago, proposed the first multiverse theory in 1895. He had a problem: the second law of thermodynamics implies that the Universe is headed towards thermal equilibrium. The Universe is about 13.8 billion years old, which seems pretty ancient, but it is nothing compared to the eternity of years ahead. The common belief in Boltzmann's day was that the Universe was past-eternal, infinitely old. And yet, the Universe is far from its final, run-down state. Why?

Boltzmann had a radical idea. He was amongst the first to realize that the second law is a *statistical* law, and to appreciate what this means: it isn't certain, but merely overwhelmingly likely, that the entropy of a closed system will increase with time. Now and again, order does spontaneously arise from disorder.

So, do you need some low-entropy energy? Just wait . . . a really long time. Boltzmann thought that perhaps, despite appearances, the universe as a whole *is* in thermal equilibrium, resembling the distant future we expect for the part of the Universe that we see around us. Most of the Universe would be lifeless, without so much as a scrap of low-entropy energy to make anything that could metabolize or reproduce. But if we wait long enough, a region of the Universe will fluctuate into a low-entropy state. Life can then ride the entropy back to equilibrium.

Think about this for a second. If this is the case, the exciting Universe we see around us is really a statistical, low-entropy fluctuation in a sea of disorder. The overall universe could have existed indefinitely – or even infinitely – before the fluctuation that became our Universe. And there could have been countless other universes born out of fluctuations before ours, and there will be countless more after we fade back into the featureless background!

Luke: Interesting, if a little gloomy. But where does the idea fail?

Geraint: Let's take a survey of all the life forms in the universe. And not just the life forms at a particular instant, but also the life forms over all of the immense cosmic history of this multiverse. As I mentioned, even if this entire multiverse structure had a lively birth, the vast majority of its time is spent in this high-entropy, dead state. Given the available eternity, we'd expect life forms in fluctuations to vastly outnumber the life forms that thrived when the universe was in its low-entropy youth. In this picture, we are, most likely, a random fluctuation in a more-or-less dead cosmos.

Luke: Still gloomy. Go on.

Geraint: Well, allow me to introduce the *Boltzmann Brains*.

Luke: Sounds like a good name for a Statistical Physics 101 study group.

Geraint: Or the nerdiest folk band ever assembled.

Let's think about these statistical fluctuations a little more. The air around us undergoes such fluctuations, but you would have to wait a very long time for the atoms to fluctuate in such a way as to assemble a teapot. We'd require a few nuclear reactions to create the required elements, for a start!

Now, think of the fluctuation we would need for an entire, liveable universe to spontaneously arise from the sea of dead photons and matter in an aging cosmos. The chances of spontaneously forming something small, like our teapot, are much, much higher than the chance of spontaneously forming a large patch of liveable universe.

So, Boltzmann's multiverse relies on extremely rare fluctuations to provide the conditions for life, and small universes are more likely to appear than large universes. Overwhelmingly more likely. If life has to wait a squillion years to get a cubic light-year of the special conditions it needs, then it is very unlikely that the cubic light-year next to ours also chose this moment to produce a low-entropy, life-permitting fluctuation. So, you get what you need and almost certainly no more.

As Arthur Eddington pointed out in the 1930s, it is extremely unlikely in Boltzmann's multiverse that we would see a universe like this one, with low-entropy pastures as far as the eye, and the telescope, can see. Boltzmann's idea doesn't explain why our Universe, our fluctuation, appears to be so large.

So, this multiverse fails spectacularly. If our Universe were like that proposed by Boltzmann, being nothing but a large fluctuation in a larger dead universe, then we'd expect to observe ourselves crammed into a small island of low entropy amidst a sea of thermal equilibrium.

Luke: This is a straightforward failed prediction. Scientific theories place their bets on the outcome of observations, and are rewarded or punished accordingly. Betting it all on a winner is better than a range of hedged bets, which is in turn better than going for broke on a loser.

Boltzmann's theory places its bets in the 'size of the universe' market. Having put almost all its money on 'as small as necessary for life', it is bankrupted when 'vastly larger than necessary for life' crosses the finish line in first place. This is a nice illustration of how multiverse theories can be tested – and fail!

So, where are these Boltzmann Brains?

Geraint: As we've seen, even though the low-entropy fluctuations are extremely rare, they are much more likely to produce a teapot than a life-bearing universe. Similarly, it is much more likely for a fluctuation to produce a fully formed brain, complete with memories of a body and life that never happened, than it is to fluctuate into a huge, life-bearing universe.

These Boltzmann Brains are self-aware, with some who think about cosmological theories and entropy, and some who spend their few conscious moments with a Taylor Swift song stuck in their head. In the long, dark, empty future of expanding universes, Boltzmann Brains come to vastly outnumber the various life forms that existed when the universe was middle aged and filled with stars and planets.

In fact, I am in danger of concluding that I am nothing but a Boltzmann Brain sitting in a long dead universe, formed with all of the memories of growing up in Wales when music was actually good, with parents, siblings, wife and children, and with clarity about General Relativity and electromagnetism, but confusion about mortgages and taxes, and with the instantaneous thought that I am sitting here talking to you. But it could all be simply an illusion in a random fluctuation that will melt back into the background!

Luke: OK, but surely no sane person believes that they are really a Boltzmann Brain?

Geraint: Certainly, but how do you know? How *could* you know? You can't step outside yourself and look around. This is what philosophers call the 'brain in a vat' thought experiment: how do I know that I'm not in the Matrix? In the end, our belief in an external world seems to be a reasonable assumption, but not much more.

Of course, for every well-formed Boltzmann Brain with your memories, there will be many that are badly formed, with a cacophony of random thoughts and memories. This would seem to make the existence of a brain that imagines a universe like this, so orderly and consistent, extremely unlikely. Nevertheless, given the eternity ahead, your brain should arise at some point.

Luke: It seems like there are two problems here that we risk running together.

Geraint: Hit it. Problem One ...

Luke: I'll call the *Boltzmann Me* problem. Let's suppose that reality is how I perceive it: I'm a human being on Earth, formed by biological evolution, living in a large, entropy-producing universe that is about 13.8 billion years old. That's me, the Real Me.

Now, leave the universe to its own fluctuations for a while. Eventually, a brain will form somewhere out in space that believes exactly the same things about itself that I believe about myself. It will have memories and 'sensory' experiences identical to mine, believing that it lives on Earth, grew up in Macksville, and all that. This is a Boltzmann Me. While it will take an awfully long time, eventually there will be many Boltzmann Me's. The one brain that correctly believes that it is the Real Me is vastly outnumbered by brains that falsely believe that they are the Real Me.

Now comes the awkward question. Which one am I? Given that they all have identical experiences, I should conclude that I am equally likely to be any of the Brains. If there are trillions of Boltzmann Me's, then the probability that I am the Real Me is less than one in a trillion.

The problem with this scenario is that it saws off the branch that it is sitting on. Remember the logic: if I understand the evolution of the Universe and entropy and all that, then I am probably a Boltzmann Brain. But if I'm a Boltzmann Brain, then my experiences are not of a real universe. They are just the fevered fancy of a fleeting fluctuation. So, I *don't* understand the evolution of the Universe and entropy and all that – I've never done any science! The argument defeats itself. The Boltzmann Me problem is a worry, but doesn't prove much.

Geraint: A Boltzmann oddity. And Problem Two . . .

Luke: I'll call the *Boltzmann Observer* problem, and this one can really sting. All this Boltzmann Brain talk might seem like a ridiculous flight of fancy. Why are these daft physicists wasting their time on such fictions? This is to miss the point. The most worrying place that Boltzmann Brains appear is *in our theories*.

If you think you know how the Universe works, then you cannot cherry-pick the nice bits of your idea. You have to take your theory seriously, actively seeking *all* its consequences. If any seem ridiculous, then you might have a problem.

Cosmologists have discovered Boltzmann Brains lurking in theories that we thought were quite un-ridiculous, like inflation. In fact, they might even be found in the predicted future of this Universe – no multiverse required. The problem is *not* that we might be Boltzmann Brains. The problem is that we aren't.

The ultimate test of a scientific theory is in predicting observations. *Any* observation can test a theory, even something as mundane as 'I am not an isolated brain surrounded by thermal equilibrium'. So, if we take Boltzmann's multiverse seriously, and rigorously derive its

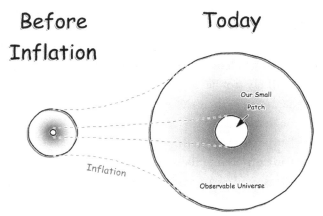

NOT TO SCALE!

FIGURE 49 Our small, local patch of space – the Solar System, say – and our large observable Universe, both before inflation (left) and today (right).

predictions, we would expect with overwhelming probability to observe that we are isolated brains surrounded by thermal equilibrium. Prediction meets observation ... and a theory goes in the dustbin.

The Boltzmann Observer problem is this: we are not Boltzmann Brains. If your theory predicts that the vast majority of observers are Boltzmann Brains, then it's in big trouble.

So, are your inflationary multiverses infested?

Geraint: Perhaps. Roger Penrose from Oxford University has argued that the modern inflationary multiverse suffers the same fate as Boltzmann's multiverse. The argument will sound familiar. The region of the Universe that we can see – the *observable* Universe – is in a low-entropy state. While it is very pretty, we don't need all this order; a much smaller patch, at most a small percentage of the total, would suffice for life on Earth. Here's a diagram (Figure 49).

Now rewind the cosmic clock, keeping a close eye on our big observable Universe and the smaller patch that sustains us. We can follow these patches back before inflation, when they were both very small regions waiting to be expanded.

Consider the small white circle on the left; call it our cosmic egg. Inside, the conditions are just right for inflation to begin, continue and end, producing our small patch on the right. The egg is in a highly ordered, low-entropy state. We know this because it evolves into our smooth neighbourhood, which from gravity's perspective is an ordered state of affairs.

Since our observable Universe is in a low-entropy state, the region around our cosmic egg must also be just right for inflation. But life doesn't need it, so why is it there? These just-right-for-inflation conditions are very rare, which means that large observable universes like ours are also very rare in the inflationary multiverse.

How rare? Penrose gives a rough calculation: for every observer who observes a smooth, orderly observable universe as big as ours, there are 10 to the power of 10^{123} who observe a smooth, orderly universe that is just 10 times smaller. That's a number too large to even pretend to comprehend. 10^{123} is trillions of trillions of trillions of times larger than the number of particles in the observable universe, and that's just the number of *digits* in Penrose's number.

If Penrose's argument is correct, then – as before – most observers in the inflationary multiverse will be Boltzmann Brains, small islands of order in a sea of chaos. And so the theory seems headed for the dustbin.

Luke: Isn't the whole point of inflation that it makes big, smooth universes? Whence this plague of small universes and their Brainy occupants?

Geraint: Think back to what inflation needs.[8] We need space to double in size at least 80 times to get a Universe as big and flat and smooth as ours. Inflation also needs a certain amount of smoothness just to get started at all. Penrose's argument suggests that most regions of the Universe that inflate won't start smooth enough and increase in size enough to become as big and smooth as ours.[9]

---

[8] Point C in the 'What inflation needs' list from Chapter 5.

[9] Other physicists have advanced arguments similar to that of Penrose (2004), that is, that inflation itself requires special initial conditions to produce a large universe, and

Luke: So, Mr Physicist, one of your precious theories has produced an absurdly low probability for something that we know is true. What do you do now? Give up on all this 'science'? Retreat in shame?

Geraint: Sarcasm noted, though we've both seen non-physicists react like this to scientific theories. Rather than concluding that one of our ideas is wrong – gasp! – apparently this shows that modern physics is a farce. As if we've never made a mistake before.

What the physicist does next is to work backwards from the ridiculous conclusion to identify the underlying assumptions. In Penrose's argument, they range from well-tested physics to reasonable extrapolation to speculation. Which of these, if replaced, would avoid the 'one in ten to the power of ten to the power of one hundred and twenty three' probability?

Penrose, for example, sees the need for principles that govern initial conditions. He argues for something called the 'Weyl Hypothesis', which implies that the beginning of a universe is always smooth and orderly.

The elephant in the room is quantum gravity. We don't have a theory that can describe the Universe's earliest stages. In particular, what – if anything – came before inflation? Is there a smallest possible patch of space? And what will quantum gravity have to say about the entropy of an expanding universe? These are difficult questions.

Luke: OK. I'll try to summarize, and we'll see if we can reach any conclusions. We can see how the multiverse is supposed to explain fine-tuning – make lots of different universes, and then life will appear in only the fine-tuned ones.

Encouragingly, we already have some of the necessary pieces. We have theories of physics and cosmology that, while not proposed for this reason, seem to be able to create a multiverse. String theory promotes the 'constants' of nature to dynamical, changeable things

indeed to occur at all. This is a technical point, so we'll include a few references for those who wish to read more. For popular-level expositions, see Carroll (2006 and 2010). For more technical details, see Page (1983), Hollands and Wald (2002), Albrecht and Sorbo (2004), Carroll and Tam (2010), and Carroll (2014). For an opposing view, see Kofman, Linde and Mukhanov (2002).

such as fields. Cosmic inflation can create universe domains that realize the different possibilities of string theory.

Another plus is that, despite appearing to be lifted from science fiction, the multiverse actually faces serious tests as a scientific theory. Most importantly, it can make predictions. In principle, we should be able to sit down with a multiverse theory and derive what we'd expect to observe. A good multiverse theory will predict something that our best current theories cannot – the values of the constants. Beginning with Weinberg's prediction of the cosmological constant, this approach has had some success.

Unsurprisingly, the multiverse can be criticized. It cannot be directly detected. If the multiverse exists, we will never get information from the other universes. They could all disappear tomorrow and we'd never know. This is an important difference between the multiverse and other kinds of unobservable entities in physics. If all the quarks in the Universe disappeared, we'd know about it. Well, probably not. We'd disappear with them. This is *not* a deal breaker – evidence is evidence, even indirect evidence – but it is reason to proceed with caution.

Also, the pieces borrowed from modern physics aren't particularly secure. While inflation has made successful predictions, its mechanism is not known. Whatever its mathematical appeal, there is zero experimental evidence for string theory and even less for the string theory landscape. The theory is so difficult that few of its predictions are secure. For example, Tom Banks of the University of California argues that the landscape is populated by different *theories*, further complicating the mix of potential universes. In the same way that, within Newton's theory of gravity, rearranging masses cannot change the strength of gravity itself (G), no physical process – including cosmic inflation – could realize the different possibilities in the string landscape. Whether Banks is correct or not, this demonstrates that disagreements between the experts run *deep*.

There is so much that we haven't discussed. There are other theories of quantum gravity, such as *loop quantum gravity*, and even inflation has competitors, such as Robert Brandenberger's 'string gas cosmology'. It might seem that physicists have a theory in every pocket. On the contrary, my impression is that our theories are but a few pebbles on the shore of an ocean of knowledge. We have barely begun to explore.

Despite these uncertainties, the Boltzmann Observer problem is formidable. The multiverse has a tightrope to walk. Too few varied universes, and it will probably fail to make a life-permitting one at all. Too many non-fine-tuned universes, on the other hand, could result in a universe filled with Boltzmann Brains. Even if your multiverse all but guarantees the existence of large, highly ordered, star- and planet-forming, biological life-permitting universes, most observers would not see them.

Geraint: The positive way to view these problems is to say that they will help us discover the real multiverse. They help us find the diamonds in the rough.

Luke: Correct, but on the other hand, if enough multiverse theories fall by the wayside, then the whole enterprise looks suspicious. We would need to … ahem … *fine-tune the multiverse.*

Geraint: Well, there's certainly more science to be done, more equations to solve, more equations to *discover*, more of the Universe to explore, and more experiments to build and analyse.

We've come a long way. The process that drives inflation and creates the multiverse doesn't actually care that, in at least one of the universes it created, conditions were right for the formation of stars, the growth of rocky planets, and descendants of east African apes who can ponder their own existence. But the fact that us hairless chimps do this is a constraint on the mechanism, a mechanism we have barely begun to unravel. It's going to take a lot of work, generating a lot of entropy through the thinking processes of cosmologists and quantum physicists, to solve this. We've barely begun!

Luke: On the other hand, we're pushing back to the beginning of the Universe. We're reaching towards the ultimate laws of nature.

We've asked more 'why?' questions than a four-year-old at the zoo. Maybe we're glimpsing the limits of what science can explain.

There is another, older answer.

Geraint: I know where this is going!

THE G WORD!

Luke: Simply put, the Universe contains life because it was made that way.

Geraint: I think I've heard this one before. By 'made' you mean literally designed and constructed?

Luke: Yes, designed and constructed. As long as people have been thinking about why the world is the way it is, it has seemed obvious to many that its various parts are put together in a purposeful way. Together, the pieces of the Universe achieve something: the Sun gives life, the Earth gives nutrients, the clouds give water, plants convert all that into food and oxygen for animals to thrive and for human beings to be born, grow, learn, love, and labour. The whole system seems well thought out, suggesting that someone planned and created it.

Geraint: Science has supplanted many of these explanations, though. We no longer believe in thunder gods – or at least the laws of electromagnetism and the natural properties of clouds have put them out of a job. We know what powers stars, we've explained the chemical basis for photosynthesis, and we know the biochemical processes that underlie life. What's left for a god, or gods, to do?

Luke: This is precisely why fine-tuning has spurred a renewed interest in this argument. Science has hierarchies of explanation. We explain the phenomena we observe in terms of various laws, and we even explain laws in terms of more fundamental laws, as when we say that chemical laws follow from the quantum mechanical laws that govern electrons orbiting nuclei in atoms.

So, if I say, 'life seems rather well put-together', and you reply, 'life depends on chemistry', the next question to ask is: 'is chemistry well put-together?' We should delve deeper into the scientific laws of our Universe. Fine-tuning suggests that, at the deepest level that physics has reached, the Universe is well put-together. The vast majority of other constants and laws don't permit anything like life.

Geraint: We just debated whether multiverse theories are part of science. We agreed that they could be: while the impossibility of actually observing another universe is reason for caution, empirical evidence can be advanced for and against a particular multiverse theory.

However, if anything isn't science, it's a god and gods. For a start, which of the many gods and goddesses revered by various people around the globe are we talking about?

Luke: The argument from design, or *teleological, argument* is part of the Western philosophical tradition, and is one of a host of arguments for the existence of God.[10] In particular, whether or not any are successful, these arguments provide an invaluable insight into exactly what this God idea is supposed to be about.

Perhaps the most famous set of arguments is the 'Five Ways' of St Thomas Aquinas, a thirteenth-century philosopher and theologian. By way of illustrating his importance, in the sixteenth century, when the Roman Catholic Church met at the epochal *Council of Trent*, three books were reverenced on the altar: the Bible, the decrees of the Popes, and Aquinas's *Summa Theologica*. The *Summa* is a text-book, meant for students with two years of Aristotelian philosophy under their belt. Before he embarks on three-and-a-half thousand pages on the attributes of God, creation, humankind, the will, the church, ethics, law and more, Aquinas spends about a page summarizing five arguments for God's existence presented at length in his previous books.

The teleological argument is Aquinas's fifth way. Before he gets to it, he presents three versions of what is known as the *cosmological*

---

[10] Nathan Schneider's enjoyable book *'God in Proof'* (2013) discusses the history of these arguments.

argument for the existence of God. In understanding this God character, we'll start there.[11]

Geraint: Cosmological, you say? Is there a connection with cosmology as we know it?

Luke: Yes, in the sense that the Greek word *cosmos* can refer to the whole Universe. The cosmological argument is motivated by a sense that the Universe needs an explanation over and above itself, that the things with which we are familiar point to a *different kind of thing*.[12] It is an attempt to tie off the loose end of a child's 'Why? But why? But why?' questions.

Philosophers have tried to identify *what* about the Universe needs this extra explanation. For Plato and Aristotle, it is 'motion' (meaning change of any kind), which points to an unmoved mover. For certain Arabic philosophers, the beginning of the Universe suggests that it has an external cause. In Aquinas's second way, it is causal power: the chain of cause and effect in which nothing is the cause of its own existence. In Aquinas's third way, it is *contingency*: the Universe could have been different, and so cannot explain why it is this way and not otherwise. For the German philosopher Gottfried Leibniz, it is existence: there must be a sufficient reason why anything exists at all. The cosmological argument, as you can see, is actually a family of arguments.

Geraint: OK. This is not quite like cosmology as I know it, but go on.

---

[11] For the curious, the fourth way is known as the argument from degrees of perfection. Of the five, it is the most likely to be misunderstood by the modern reader, who has probably not studied Aristotle and has absorbed a very different worldview (and vocabulary) from the twenty-first century. The interested reader can get a useful crash course from Feser (2009), and a modernized version from Ward (2009). If you want to read Aquinas, Peter Kreeft's *A Summa of the Summa* (1990) extracts and explains some of the important bits.

[12] Saying *anything* about God will stir someone to say, 'That's not what *I'm* talking about'. For example, Aquinas would object to talk of God as a kind of thing, or even a being at all. Existence is not something God *has*, but rather part of what God *is*: 'God is the principle of all being'. If you'll pardon my inexpertise, I'll try to say something that someone might agree with.

Luke: Let's start with an *analogy*. The things around you, if you lifted them up and dropped them, would fall to the floor. (If you have a helium balloon, drop it in a vacuum.) And yet, most things around you are not falling. Is it because of the floor, or the foundations, or the soil, or the rocks? Seemingly not – dig them up, hold them aloft and they'll fall too. If everything is a 'falling' kind of thing, then why isn't everything falling?

Here are three options:

A. It's dirt and rocks all the way down, each layer supporting the one above and supported by the one below.
B. A brute fact supporter, a magical floating layer. A layer of stuff down there, despite being made of the kind of stuff that would ordinarily fall, doesn't. For no reason at all. The question 'why is the magic layer not falling?' has no answer.
C. A self-supporter. There is something down there that, somehow, supports itself. The question 'why is this layer not falling?' is answered *internally*, entirely in terms of the kind of thing the layer is.

It seems to me that A is no explanation at all, since all the layers could be falling together. We still don't know why everything isn't falling. B would be disappointing, and almost unintelligible. How odd for the stuff of the Universe to be so orderly and intelligible, conforming to cause – supported from below – and effect – not falling – only to be unexplainable at its deepest level! What an anticlimax!

C is in fact correct, and the most satisfying answer. The *core of the Earth* is supported by its internal pressure and contains the Earth's centre of gravity. That *kind of thing* can support itself against its own gravity, and can support other things.

Here's the logic of the argument. In order to make sense of why all the things that could be falling are not falling, we need to posit a different kind of thing, an unsupported supporter or self-supporter. We need to take the series of supporters – I'm supported by my chair, supported by the floor, by the foundations, by the ground and so on – and find an appropriate, rather than arbitrary, stopping place.

Take that argument, replace 'is supported' with 'exists' and you've got a rough version of the cosmological argument. In order to make sense of why all the things that could have failed to exist – *contingent* things – do in fact exist, we must posit the existence of a different kind of thing, an uncaused cause, an uncreated creator, a *necessary* being. This is an attempt to tie off the loose ends of explanations. We require God to be the kind of being that must exist, whose reason for existence is found within its own nature.

Geraint: Why couldn't the Universe be that kind of thing?

Luke: Good question, but before I get to the answer ...

Almost certainly, any objection to these arguments that you can think of has been raised and debated for millennia. For a few thousand years, some of the finest minds in the Western world (and beyond) have wrestled with the issues raised by God, existence, causality, time, reason, logic, explanation, contingency, necessity, and more. They have a 2,500-year head start. You might not agree with what they said, but you should at least find out what it is. The thing to do with your objection to these arguments is to *dig deeper*.[13]

A million feeble objections pass harmlessly by the cosmological argument. The classic is, if everything has a cause, then what caused God? Or, more succinctly, who made God? This is like objecting, what's holding up the core of the Earth? The objector reveals not a flaw, but their own lack of comprehension. To explain 'things that could fall but in fact don't', we need a new kind of thing, a thing that cannot fall. To explain 'things that could have failed to exist but in fact do exist', we need a new kind of thing, a thing that cannot fail to exist. The whole point of the argument is to *deny* that everything has a cause. The argument raises plenty of puzzles, but 'who made God?' is not one of them.

[13] If you need some help digging, Oppy (2009) and Sobel (2009) are highly regarded critiques of these arguments. For defences, see Swinburne (2004), and *The Blackwell Companion to Natural Theology* (Craig and Moreland, 2009).

Geraint: But if these ancient philosophers all believed in God, how much scrutiny did these arguments really get?

Luke: Quite a bit, actually. To believe in God is not to believe that every argument for God is a good one. Aquinas didn't think that you could prove that the Universe has a beginning. Many Muslim and Christian philosophers viewed Aristotle's ideas with suspicion. Blaise Pascal, seventeenth-century scientist and philosopher, thought that all the 'proofs' failed and so we must *wager* on whether God exists.

Further, the most famous critics of the *cosmological argument* are the eighteenth-century philosophers David Hume and Immanuel Kant. Their legendary critiques are a couple of centuries old, so the argument has been prodded from every side. The versions of the argument that are defended today may not succeed, but they do not wither at the touch of a first-year philosophy student.

Anyway, back to your question: why not suppose that the Universe itself, or something in the Universe, is the uncaused cause or necessary being? The short answer is: the physical universe just doesn't seem to be that kind of thing. There is nothing necessary about how it is, or that it is, or how it behaves. This is why science needs observations; we can't figure out the Universe from our armchairs. We need to go outside and look.

Geraint: We can't recount a few millennia of debate here, so let's remember why we took this detour. What does this argument tell us about what this particular *God* is supposed to be like?

Luke: The main point is that God is a *necessary being*. Here's a useful summary:

> If God exists, he cannot be the kind of being that just happens to exist; that is, his existence cannot be a state of affairs that obtains merely by chance . . . God cannot be a perishable being. . . . Finally, it cannot be true that God might not have existed. His existence cannot simply be a *contingent* matter, as it is with such objects as human beings. Our existence, unfortunately, is not guaranteed. But God cannot be like that. God is thought to be a self-existent being, a

being who has the reason for his existence within his own nature, and so could not be merely contingent. God's existence must in some sense be *necessary*, for this seems to be part of the very concept of God.[14]

To end the 'but why?' questions, yet another contingent being will not suffice. So we look for a different *kind* of explanation. We need a being that does not look outside itself for the reason and cause of its existence.

Philosophers disagree about the type of necessity that God is supposed to have. The strongest type – logical necessity – doesn't seem to apply, since there is no strict logical contradiction in the sentence 'God does not exist'. Some theists argue that God is merely *factually necessary*: God does not depend on any other being to exist, and all other beings depend on God to exist. In between is *metaphysical necessity*: there is no possible world in which God does not exist, even though God's non-existence is not a strict logical contradiction. Whatever the case, God is not just another item on reality's shopping list.

Remember: we're trying to understand the *concept* of God. Asking, 'why assume that God is a necessary being?' is like asking, 'why assume that unicorns and rhinoceroses have one horn?' Having one horn is part of the concept of a unicorn or rhinoceros. The question to ask is: does anything in reality match our concept? Does God, a necessary being, exist?

Geraint: Why the focus on this particular attribute?

Luke: Because it clarifies God's relationship with the Universe and the laws of nature. *Theism* – the belief that God exists – is not the idea that the Universe is a self-contained system that explains and runs itself ... but also there is a magical sky fairy roaming around outside looking for a job. The Universe is entirely dependent on God for its existence, moment by moment. God is the cause of the existence of every contingent thing, and so the laws of nature *simply are*

---

[14] White (1979).

the way that God has freely chosen to run the Universe. They are not necessary principles over and above God, or separate to God.

The relationship of God to the Universe is something like that of an author to a book. We won't find J. K. Rowling in Hogwarts or Shakespeare in fair Verona. We can't put the author out of a job by discovering a new character, or deciphering the plot, or finding the first page of the book, which reveals how the story started. This is not a hastily revised 'modern' God, retreating in the face of science. It predates the scientific revolution by a few thousand years, and for the most part was the worldview of the makers of the scientific revolution.[15]

Moreover, faced with science's *in principle* silence regarding questions like 'why does anything exist?', 'why does the Universe obey mathematical laws?', 'what is it that breathes fire into the equations and makes a universe for them to describe?', as Stephen Hawking famously asked (1988, p. 174), and 'why is the Universe as it is and not otherwise?', a beardy man in the clouds will not do. Santa Claus will not do. Sky fairies will not do. The Flying Spaghetti Monster will not do. The many gods of ancient Greece were at the disposal of Plato and Aristotle, and yet they saw the need for something else, something that wasn't just another thing, or even *just another god.*

Here is the crucial point. If God does *not* exist, it isn't because we found a replacement. It is not because of anything science has discovered. It is because reality didn't need that kind of explanation after all, perhaps because that kind of explanation doesn't really make sense.

---

[15] I'm not a historian, so let me quote one. Margaret J. Osler (2010) says, 'The entire enterprise of studying the natural world was embedded in a theological framework that emphasized divine creation, design, and providence. These themes are prominent in the writings of almost all the major seventeenth-century natural philosophers.' Of course, one mustn't claim too much. Modern science owes much to the efforts of Arabic, Greek, Roman and Babylonian thinkers, amongst others, as well as commercial, civil and military interests in new technology.

Geraint: I've got a whole bunch of questions. Let's rattle through them.

Honestly, the idea of a 'necessary being' is a pretty baffling. I know what a necessary *truth* is, like 2 + 2 = 4 or all bachelors are unmarried. But consider these examples more closely. Who says 2 + 2 = 4? The axioms of arithmetic. Who says all bachelors are unmarried? That's what the word *bachelor* means. Who says God exists? Ummm ... God? God's nature? The definition of God? Metaphysical necessity, whatever that is?

Luke: These are the kinds of issues that philosophers and theologians have debated for a few millennia. Even theists don't agree amongst themselves.

A necessary being's reason for existence is found in its own nature. This is how God is supposed to tie off the string of 'and why does *that* exist?' questions. Perhaps we can think of it like this. The statements '2 + 2 = 4' and 'my name is Luke' are both true, but there is something different about the *way* in which they are true. The first *must* be true, whereas the second just *happens* to be true. The first is *true no matter what*, whereas the second is *dependent* on my parent's choice, which could have been otherwise. *Modal logic* is the extension of basic 'true or false' logic to include these different types or *modes* of truth. In the vernacular, '2 + 2 = 4' is necessarily true, while 'my name is Luke' is only contingently true.

So, just as there are different modes of truth, there may be different modes of existence. God has a necessary mode of existence. The attraction of this idea is that it would explain why anything exists at all, and why there are objects with a contingent mode of existence, that is, things like us and planets and the Universe that exist but don't exist necessarily.[16]

Geraint: Next question – is God complex? In particular, is God more complicated than the complicated things being explained?

[16] Modal logic has developed significantly in the past century. See the forthcoming book 'Necessary Beings' by Alexander Pruss and Joshua Rasmussen for a discussion of how these developments inform the debate over the existence of God.

Luke: That's a very important, and common, question, because any argument for the existence of God sinks rapidly if God is complex, *and* this complexity makes God's existence improbable.

Richard Dawkins presents this objection in Chapter 4 of *The God Delusion*. His version is a rough sketch, and so we have to fill in a few details; I *think* that the argument is supposed to go like this.

(a) Organized, complex things are improbable.
(b) A designer is more organized and more complex than the thing designed.
(c) Thus, a designer is more improbable than the thing designed.
(d) Thus, it is extremely improbable that God (the ultimate designer) exists. It is more likely that the Universe just exists.

Here are two things to say in response. Firstly, is God complex? The Universe is a complicated place, so there must be *some* place in our ideas for complication. The trick is to get the complications to flow *from* our ideas, not to be fed *into* our ideas. An idea is not penalized by its complicated *consequences*. And God is usually defined with a few statements: a necessary, free, omnipotent being. Some would argue that still shorter definitions are possible: God is a perfect being, or the greatest conceivable being.

Secondly, I think that the argument commits the fallacy of *equivocation*; that is, one of the words has changed its meaning over the course of the argument. Dawkins says:

> The same kind of intuitive calculation lies behind the claim that the vertebrate eye is too improbable to have arisen by chance (in how many ways could the bits of an eye have been arranged, and how many of them would see?) and it lies behind my similar claim about God.[17]

To be more precise, then, we should rewrite premise (a) as 'Organized, complex things are unlikely to have been assembled by chance'. But if that is what 'improbable' means in the argument, then

[17] Dawkins (1995).

its conclusion should read, 'It is extremely improbable that God was assembled by chance', a conclusion with which the theist can heartily agree; it does not follow that God's *existence* is improbable.

Geraint: God is supposed to exist necessarily, but what about a universe in which nothing exists? I can imagine such a universe, and there's no God there.

Luke: Certainly, there is no logical contradiction in the statement 'nothing exists'. But is what you can *imagine* a guide to what is absolutely possible? And are you really imagining *nothing*, or just a dark, empty space?

Geraint: Saying that God's existence is part of what God is seems to confuse *what* a thing is with *whether* a thing is. Surely, 'what is God?' and 'does God exist?' are separate questions.

Luke: Even for those who believe that God's existence is absolutely necessary, 'what is God?' and 'does God exist?' are not the same question. It makes sense to ask: does a necessary being exist?

The stakes, however, have been raised. Such a being can't just happen to exist. God's existence is either necessary or impossible.

Geraint: If we instead go with *factually* necessary – as you mentioned above – then God doesn't seem to answer the question of why anything exists. God is the ultimate contingent thing, but a contingent thing nonetheless.

Luke: That's correct: a factually necessary God is metaphysically contingent, and so unable to explain why anything exists. To some, this is sufficient reason to discard the idea. To others, the cosmological argument claims instead that God is the simplest explanation of the totality of contingent things – they all are ultimately caused by one all-powerful, free entity.

I think that the worry behind your questions is that this *God* idea seems to be cheating. How can a thing, just because of its own nature, explain its own existence? I think many theistic philosophers would agree that a necessary being is a very strange idea. God is totally unlike anything with which we are familiar. But that's the point. The core of the Earth isn't like the floor of this house. The cosmological argument looks for something that can explain *existence itself*; the familiar stuff of everyday life just won't do. The contingent things around us send us in search of an explanation, but cannot provide that explanation. We'll just have to get used to the strangeness of whatever explains existence itself, because, in my opinion, the alternative is worse.

Geraint: And what is the alternative?

Luke: Naturalism.[18] Physical stuff is the only stuff. The ultimate laws of nature are the ultimate principles of all reality. The Universe is absolutely contingent and ultimately unexplained. Of all the things that could have existed and events that could have happened, why this stuff and these events? The naturalist says that this question has no answer. Not that we don't know the answer and need more thinking or more evidence. The question is *unanswerable, in principle*.

Geraint: We've found the simplest option, surely. No need for unknown, bewildering kinds of beings. As in science, we'll stay close to the empirical evidence and simply say, 'that's just how things are'. To quote Sean Carroll, 'there is a chain of explanations concerning things that happen in the universe,

---

[18] Well, there are always more alternatives. Hume argued that an infinite sequence of contingent things, each explained by the one before and explaining the one after, is an adequate and non-necessary explanation. This is like Option A: Earth all the way down. The infinite sequence of things could have been different, and could have not existed at all, so we've still got questions.

We could say, with the seventeenth-century Dutch philosopher Spinoza, that the Universe – despite appearances – is not contingent. Everything happening around us must exist and every event must happen, necessarily. Alternatively, there could be a necessary being that is more like a mechanism than a person. This would seem to make the Universe necessary too, and in any case handles fine-tuning in the same way as naturalism. Then there's John Leslie's *Axiarchism*: contingent beings exist because they *should*. Ethical requirements are creative. I suppose it would be *good* if that were true . . .

I am not a philosopher, and cannot adequately describe let alone critique all the options. A popular account is given in Holt (2012) and a more sustained discussion in O'Connor (2008).

which ultimately reaches to the fundamental laws of nature and stops.'[19]

The whole point of the ultimate laws of nature is that they explain the empirical facts of science. This is in principle, of course, because in practice solving the equations is often beyond us. Nevertheless, this is not insignificant. With the fundamental laws of nature in hand, we have a simple, precise mathematical theory that explains all the data gathered by science. If there's ever a time to high-five and head for the pub, this is it.

Luke: Keep in mind what's doing the explaining. Using our best understanding of the scientific laws of the Universe, we can understand the physical phenomena we see around us. Schematically,

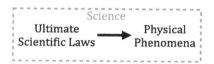

However, science, *as science*, is agnostic about whether anything explains the ultimate laws of nature.

Naturalism, in contrast, is not agnostic. Rather, it is the claim that there is nothing that explains the ultimate laws of nature. There are no arrows into the 'science' box,

In the diagram, 'nothing', as is customary in English, means 'not anything', rather than a special type of something or even just a little bit of something. Why some cosmologists fail English 101 over this word is beyond me.

[19] edge.org/discourse/science_faith.html

In contrast, theism postulates that God created and sustains the Universe in a way that is consistent, rational and discoverable. Discovering those laws allows us to understand and even predict physical phenomena.[20] In a diagram,

Again, this is not a post-scientific-revolution, revisionist God being pushed further and further out of the picture by progress. Here's St Augustine, circa AD 415: 'The ordinary course of nature in the whole of creation has certain natural laws ... determining for each thing what it can do or not do. ... Over this whole movement and course of nature there is the power of the Creator.' And Aquinas, in the thirteenth century: 'There are certain intermediaries of God's providence; for He governs things inferior by superior, not on account of any defect in His power, but by reason of the abundance of His goodness; so that the dignity of causality is imparted even to creatures.'[21]

God is not presented here as a competitor to any scientific theory. Theism's rival is naturalism, not science, and theism offers an explanation where naturalism offers *none*. If naturalism turns out to be more likely than theism, it will be because theism unwisely tried to explain the unexplainable, not because naturalism offered a better, or indeed any, explanation of the ultimate laws of nature. And certainly not because of any scientific theory.

So, the success of science looks the same on naturalism and theism. The ultimate laws of nature are just as contingent on both,

[20] The diagram is not supposed to represent *deism*, as the relation in the diagram holds at every place and time in the Universe.

[21] Augustine (1982, p. 92). Aquinas quoted in Kreeft (1990), Question 22.

given the great mathematical morass of possible laws. In physics, we solve equations and compare to some experimental findings, all without needing any opinion on what might be deeper than those laws. Theism doesn't tell us what the ultimate laws of nature are, so we need science. But naturalism doesn't tell us either – just telling me that there is nothing but physical stuff doesn't tell me what that physical stuff is and does. *The success of science is not the success of naturalism. The predictions of naturalism are not the predictions of science.*

Geraint: So how do we decide between naturalism and theism?

Luke: In the same way as every other idea. We think about how likely it is intrinsically. And we walk around in its shoes for a while, seeing if anything around us looks surprising or unexpected. To repeat a previous example, if we put on the 'burglar guessed the code to the safe' shoes, then we should be flabbergasted that the burglar correctly entered the 12-digit code on the first attempt. Sometimes surprising things happen, of course. But perhaps we should try the 'it was an inside job' shoes.

So, let's put on the naturalism shoes and walk around. Have a look at *this* universe. Surprised? Well, if there is a universe at all, then it has to be *some* way. Why not this way? Right?

Geraint: Ah . . . '*this* vs. *some*' is a red flag to the Bayesian.[22] One of the core principles of Bayesian probability theory is that information must not be arbitrarily ignored. All evidence must be taken into account.

For example, suppose Keith wins the lottery, choosing the correct numbers out of 10 million possibilities. The probability of Keith winning fairly is 1 in 10 million! Did Keith cheat?[23] You might be tempted to say that we should be calculating the probability of *some* person winning the lottery, not the probability of *this* person. This will not do, because it tacitly assumes that Keith is just some typical lottery player – but that is precisely the idea that we want to test. To calculate the *some person*

---

[22] Recall our discussion of Bayesian probability theory from near the end of Chapter 7.

[23] This lottery example is due to Brendon Brewer of the University of Auckland.

probability, we have to throw away what we know about Keith, and some of those things might be significant. Perhaps Keith is a known fraudster, knows the Lottery Commissioner, and was seen tampering with the machine.

Luke: Right. So we're thinking about how likely *this* Universe is on naturalism. It seems to me that while no particular universe is *surprising* on naturalism, there are a lot of possible universes – think of all the changes we've seen so far, and all the ones we haven't even thought of!

Given naturalism, there is no expectation that any possible physical universe is more likely to exist than any other, because there is nothing above the ultimate laws of nature, nothing to give us a clue that would nudge us towards any particular possibility. It's not merely that we don't know these reasons. On naturalism, there simply are no facts that explain why contingent reality is this way rather than some other way.

In probability, we call such theories *non-informative*. It's not a pejorative term; probability is about putting a number on our degree of certainty given what we know, and sometimes we don't know anything relevant. Non-informative theories tend to postulate very little, so appear to be simple. But their likelihood is at the mercy of the set of possibilities. If this set is very large, then the likelihood will be spread very thin, with the probability of any particular possibility being very small. To illustrate, the more digits required to open a safe, the more possibilities there are, and the less likely it is that an uninformed burglar will simply guess the combination.

And, as we noted above, a small likelihood is an opportunity, a chance to consider changing shoes. If another idea makes this Universe appear more likely – without being ad hoc or jerry-rigged or otherwise intrinsically implausible – then it is more likely to be true.

Geraint: I can see where this is headed. If God exists and installs the laws of nature, then – so long as something about this Universe takes God's fancy – this Universe will appear more likely.

The problem, though, is that we cannot write down equations to predict what God would do. Where are you going to get these probabilities?

Luke: A good question. God cannot be a part of a scientific theory because God is not a physical thing. But the same is true of naturalism; it isn't a scientific theory either. There are no equations of naturalism. The equations of physics look the same on both theories.

We're going to need to use our intuition here, rather than expect a precise calculation of these probabilities. The alternative is to just give up on trying to work out whether these ideas are true.

Geraint: Alright! I'm bracing myself. Here's what we've all been waiting for . . . bring on the fine-tuning.

Luke: Fine-tuning shows us a feature of the laws of nature – as we know them – that seems surprising, something about the equation that cries out for further explanation. Our Universe has the very rare talent of providing for the existence of intelligent, moral agents: life forms who are conscious and free, who can live, learn and love, who can think about mathematics and music, who can investigate their universe, who can make significant moral decisions, choosing to do good or harm to one other, who can communicate, cooperate and develop a moral character.

Geraint: But how does theism make any particular universe more likely? God is supposed to be omnipotent – to be able to do anything that doesn't involve a contradiction. If God can do everything, then God could do anything. There's nothing to stop an all-powerful being from creating whatever they please. An explanation that doesn't lead us to expect any *particular* outcome is as bad as no explanation at all. Theism is just as bad as naturalism.

Luke: Let's think about it. I can't play the drums – this is relevant, so stay with me. If you place me in a room with a drum-kit and wonder what noises you might expect to hear, my incompetence limits my options. Of all the noises that could come through the door, don't expect those of a competent drummer. On the other hand, if you invite rock legend Dave Grohl[24] – Nirvana, Foo Fighters, Queens of the Stone Age – into the room, then there is a much wider range of possible sounds. With a wider range of ability comes a wider range of possibilities, and so less reason to expect any particular possibility.

But think a bit harder about the example. Grohl isn't just more able. He's also a musician. He *could* produce a wider range of sounds, but there are certain sounds that he finds more desirable: musical sounds. He is more likely to launch into 'No One Knows' than flail like an amateur.

Here's how the argument goes. You are correct to say that God's omnipotence doesn't get us anywhere. But God is also supposed to be perfectly good. God isn't some out-of-control universe factory, indiscriminately spraying reality in all directions. We expect God to create universes that have morally desirable qualities.

Geraint: How do we know what God would or wouldn't find morally desirable? It's a bit presumptuous to suppose that we can read the mind of a morally perfect, omniscient being.

Luke: I'll try to summarize the detailed case made by Richard Swinburne in his book *The Existence of God*. The short story is that we have moral knowledge. It isn't perfect – like our knowledge of the physical reality, it is piecemeal and noisy. But, the person who believes that love is better than hate and that human beings have dignity and value believes something true, and believes it with some justification.

Here's an analogy. If we discover a mathematical theorem, then we know something that a perfect mathematician would know. Similarly, if we believe that we have discovered moral truths, then we have a glimpse of the moral principles that a perfect moral agent would have.

---

[24] We do hope Dave will read this book!

In fact, it seems like we don't need much speculation here at all. Love needs a lover. A moral universe needs *moral agents*, that is, creatures with free will, who can recognize and appreciate good things, do good actions, interact with, influence and be responsible for each other, investigate their surroundings, and learn the goodness of beauty, love, knowledge and truth. A universe capable of producing and sustaining such creatures is a universe with moral worth, one that God might create.

This 'might', it seems, is enough. Recall the card game discussed earlier, where Bob has dealt himself five flushes. Jane looks sceptical, but Bob counters: 'you don't know me! I'm a free agent; I can do as I please. It is presumptuous to assume that I would cheat.' But Jane remains sceptical, and for good reason. The probability of five royal flushes is one in a hundred billion billion billion. Thus, Jane needn't assume that Bob would cheat, or even is likely to cheat. She need only assume that, before they started playing, the probability of Bob cheating isn't comparable to one in a hundred billion billion billion. Only an extraordinarily strong presumption of Bob's *innocence* would counter Jane's accusation. Similarly, only an extraordinarily strong presumption *against* the idea that God would want to create a universe with embodied moral agents will affect our conclusions.

Here's the punch line. As we have seen, there are vast wastelands of unliveable but possible universes. Naturalism, with no further information, hedges its bets by spreading its likelihood over all the possibilities. This leaves naturalism at the mercy of the set of all the possible ways that concrete reality could have been – a decidedly unenviable position. It is just about the worst possible non-informative hypothesis, since it is at the mercy of the set of every possible way that physical reality could be.

Theism, by contrast, offers an explanation where naturalism and science can offer none. It has a hot tip, betting heavily on the tiny subset of universes that permit embodied moral agents. It wins more when this Universe turns up.[25]

Geraint: Wait a minute ... God doesn't play dice, as Einstein said (though in a different context). Why is God betting on universes?

Luke: OK, let's be clear about that betting analogy. The probabilities we're talking about are *credences*, not *chances*, to borrow some terminology from Chapter 7. We're not saying that God's creation of the Universe was chancy. Similarly, in discussing the probability of this Universe on naturalism, we're not assuming that there is some haphazard process that churns out naturalistic universes. These aren't the objective chances we find within physical theories, which describe the chancy properties of physical things.

These probabilities are Bayesian. They are about what we know. They are in *our* heads. We want to know whether these ideas are true, and so we are trying to think through their implications. Bayesian probabilities are a tool to help us. It is *theism* that places its bets, not God.

Geraint: But we've still got a picture of God as wandering into the library of possible universes, and searching high and low for a nice one. This sounds a bit like a creation of the late Terry Pratchett:[26]

---

[25] As with the cosmological argument, there is a family of design arguments. This is just one version.

A different, more famous version is due to William Paley (1743–1805), who asked: if you were walking on a heath and found a pocket watch, by what principles would you reach the conclusion that it was designed rather than an accident? Paley invites us to apply the same principles to the eye, muscles, skeleton, circulatory system and, intriguingly, the form of the law of gravity. Note that this need *not* be an argument from analogy. When applied to the biological world, it is a 'God of the gaps' argument, since God is competing directly with scientific theories. The more modern 'Intelligent Design' movement tries to argue that what we know about natural causes shows that they cannot fill a particular gap (e.g. the bacterial flagellum, the origin of life, the origin of biological information). The argument from fine-tuning is (or at least should be) different, as it does not compete with any scientific theory.

Aquinas's fifth way is different again to Paley's argument. For Aquinas, *any* causal regularity shows that natural things – not just eyes but planets and clouds and electrons – have an inherent tendency to produce some end, some effect. He concludes, 'Whatever lacks intelligence cannot move towards an end, unless it be directed by some being endowed with knowledge and intelligence; as the arrow is shot to its mark by the archer.' See Kreeft (1990) and Feser (2009) for more details.

[26] From *Soul Music* (1994, p. 13).

Gods play games with the fates of men. But first they have to get all the pieces on the board, and look all over the place for the dice.

Luke: Just as the probabilities are all in our heads, so are the possibilities. It is *we* who are faced with a set of possibilities as we walk around in the shoes of theism and naturalism. We could call them *conceptual* possibilities: *for all we know*, these other universes are possible.

Put on the naturalism shoes. Apart from self-consistency, there is no restriction on what physical universe we would expect on naturalism, because there are no principles of concrete reality over and above the laws of nature. So the set of *conceptual* possibilities is enormous: *for all we know*, any self-consistent universe could have existed on naturalism.

Now put on the theist shoes. *For all we know*, what universes are possible? This is not asking, what universes are possible for God? Unless you are God, you don't have this information so you can't take it into account. Remember that reasoning is about thinking as hard as you can with the information you have. So, in fact, the set of *conceptual* possibilities is roughly the same as before: *for all we know*, any self-consistent universe could have existed on theism.

It is the *probabilities* that they place on that set of possibilities that separates the two views.

Geraint: Why would God need to fine-tune the Universe at all? Couldn't God just create life in any old universe?

Luke: On theism, the laws of nature *just are* a description of how God runs the Universe. So it makes no sense to say that God could make a consistent exception to the laws of nature, since a consistent exception is itself a rule! I'm reminded of Lewis Carroll's *Through the Looking Glass*, in which the White Queen explains to Alice that,

> 'The rule is, jam to-morrow and jam yesterday – but never jam to-day.'

'It must come sometimes to 'jam to-day',' Alice objected.

'No, it can't,' said the Queen. 'It's jam every other day: to-day isn't any other day, you know.'[27]

Further, it seems a bit thoughtless for God to make any old universe and then have to constantly poke it to make it life permitting. A universe with the right laws in the first place is simpler, and easier to investigate scientifically.

I can, however, see a problem with the argument. I'll recap, to make it more obvious.

1. Naturalism is non-informative with respect to the ultimate laws of nature.
2. Theism prefers ultimate laws of nature that permit the existence of moral agents, such as intelligent life forms.
3. The laws and constants of nature as we know them are fine-tuned – vanishingly few will produce intelligent life.
4. Thus, the probability of this (kind of) universe is much greater on theism than naturalism.

Geraint: Right. There could be a rather large gap between the ultimate laws of nature in Steps 1 and 2 and the laws-as-we-know-them in Step 3. Physicists have made great progress in broadening and refining our understanding of how nature works, from Aristotle to Newton to Einstein,[28] to the many twentieth-century scientists who crafted quantum physics. Physics and cosmology as they stand at the beginning of the twenty-first century are almost certainly not the last word on the inner workings of the Universe. But it is the *ultimate* laws of nature about which naturalism is non-informative. We can't vary the *ultimate* laws and constants of nature to discover whether they are fine-tuned – we don't know what they are!

It seems like the argument is in a bind. If the theist claims that God fine-tuned the laws and constants of nature as we know them, then they are headed for a confrontation with science, not just

---

[27] Carroll (1871), Chapter 5.

[28] If you doubt Aristotle's place as a great physicist alongside Newton and Einstein, then read Rovelli (2015).

naturalism. The fine-tuning argument becomes another 'God of the gaps' exercise. Deeper scientific theories may explain the value of the constants, and so – yet again – remove the need for a divine explanation.

Luke: This is a worry, and this gap between the laws-as-we-know-them and the ultimate laws of nature is often glossed over. However, I don't think that it sinks the argument, for a few reasons.

The best we can do is to investigate the laws of nature as we know them, as well as plausible deeper physics. We have found fine-tuning as deep as we can go. Further, we have found that fine-tuning follows us down. It shows no sign of disappearing on deeper levels. We can explain the proton mass in terms of, amongst other things, the quark masses; those in terms of the properties of the Higgs field; those, perhaps, in terms of supersymmetry. Anthropic constraints appear at all levels. Similarly, life's requirements for the initial expansion rate, density and perturbations of the Universe can be translated into requirements on the properties of the inflaton.

Secondly, unless the ultimate laws of nature are drastically different from all the laws we are familiar with, there will still be initial conditions to vary, even if there are no constants. In string theory, for example, this is plenty of freedom – the constants of our current theories become initial conditions, but remain just as free. And we are always free to vary any and all parts of the mathematical framework, such as the form of the equation.

Thirdly, universes described by the laws-as-we-know-them are still possible universes. There is no logical or mathematical contradiction in their description – if there were, we'd discard them. They represent a chunk of possible-physics-space, and in this space life is fine-tuned. As we noted previously, even if Einstein's dream of a free-parameter-less theory were true, it *wouldn't* show that the Universe couldn't have been otherwise.

Further, it is debatable whether *on naturalism* we should expect ultimate laws of nature with no free parameters. If we're just accepting contingent facts, why not a few contingent numbers? Why expect simple theories at all, on naturalism?[29] That being the case, we *have* investigated some *possible* ultimate laws of nature, even if they are unlikely to be the ultimate laws of this Universe. And that, I think, is enough to get the argument up and running.

Geraint: Here's a big problem, in my opinion. I can imagine better universes than this one. I would expect a morally perfect being to create a morally perfect universe, or at least the best universe available. I would expect God to look at this possible universe, notice the cancer and death and hatred and wars and suffering and pain, and promptly put it back on the shelf. God would prefer a universe with moral agents, maybe, but not this mess.

The evil and suffering are very surprising if we've got the theism shoes on. They seem less surprising on naturalism, where any old universe could turn up. So it looks like this universe is not very likely on theism after all.

Luke: That's the most famous argument against the existence of God: the problem of evil and suffering. The Scottish philosopher David Hume noted, 'Epicurus's old questions are yet unanswered. Is he willing to prevent evil, but not able? Then he is impotent. Is he able, but not willing? Then he is malevolent. Is he both able and willing? Whence then is evil?'[30]

Oceans of ink have been spilt over this problem. It was debated by the ancient Greeks, compellingly raised but not directly answered in the *Book of Job*, and continues to furrow

---

[29] Is Occam's razor – expect simple theories to be true – a general principle of reasoning that should be obeyed in all possible worlds, or an empirical rule of thumb that works in our Universe? On the one hand, multiplying hypotheses would multiply probabilities, giving still smaller probabilities for complex theories. On the other hand, I think I can imagine worlds in which Occam's razor is bad advice. If the world were totally chaotic, then expecting simplicity would be naïve and usually wrong. Add this to the long list of issues raised but not answered in this chapter.

[30] Hume (1779), Part 10. Whether Hume's quotation of Epicurus (341–270 BC) is accurate is a matter of some dispute.

brows today. I'll say a few things, but we can't even begin to scratch the surface.

The basic theistic reply comes from Augustine: 'God judged it better to bring good out of evil than not to permit any evil to exist.'[31] Is that at all plausible? What good is worth the evil we see? A common suggestion is free will: a dangerous gift, but necessary for love, or indeed any morally meaningful actions. And then we can debate whether there are better worlds than this one with as much freedom of the will, and so on and so on.

Let's stick close to our topic. Notice that the design argument and the problem of evil are two sides of the same coin. Both ask: would a morally perfect being want to create a Universe like this? It is useful, then, to see if a given defence of God against the problem of evil also dismantles the design argument. For example, if the theist objects that we are limited in knowledge, space and time, and so cannot judge whether good may come from some perceived evil, then the naturalist can say that we cannot judge whether evil may come from the perceived good in this universe. So, perhaps we don't know whether God would want this universe after all.

However, remember that we are walking in the theist's shoes, seeing what the world looks like. *On theism*, good coming from apparent evil is much more likely than evil coming from apparent good. There is a good God behind the scenes, directing the Universe to good ends.

Ultimately, the theist can simply choose to bite the bullet. Yes, the evil in our Universe is surprising and I don't know why it's there and why there is so much. But I'm still not as surprised as the naturalist. The typical naturalistic universe won't contain moral agents at all, and so will contain neither evil nor good. So naturalism has a problem of evil and a problem of good – both are surprising.

Geraint: Well, I'm sure you'll agree that we won't settle these debates today. But I think I can summarize your thoughts on this. To

---

[31] *Enchiridion*, Chapter VIII.

you, the apparent fine-tuning of the properties of the Universe, the properties that allow you to exist as a living, thinking, active creature, is not an accident, not some random role of the dice in an inflating cosmology. To you, the conditions were chosen; the dials were explicitly set to allow your existence.

This universe contains good things, like free moral agents and all that they can do and learn and appreciate. The presence of these qualities is not accidental, but reflects the intent of the creator, the person who set the dials. How's that?

Luke: That's about right. But I feel that you don't buy the argument.

Geraint: Alas, I don't. I think that moral beliefs have arisen through our evolution and allowed us to survive and thrive in communities and clans. Anyway, all people appear to be a little amoral sometimes!

Luke: Don't confuse the question of how we got our moral beliefs with the question of what they are *about*. I got my eyeballs from evolutionary processes as well, but I believe them when they tell me that there's a tree over there. And certainly, knowing what is right to do is no guarantee of doing what is right. We aren't perfectly moral, but I think that we know enough to be able to think rationally about the idea of a morally perfect, necessary being.

Geraint: Still, the argument seems to be explaining a little by supposing a lot. There is another way that the Universe may have been fine-tuned by a creator, but without all the superlatives and perfections and omni-this and omni-that.

Luke: Go on.

THE SIMULATED UNIVERSE

Geraint: Maybe the Universe appears to be fine-tuned as it was programmed to be that way. And by programmed, I mean in the same way that scientists explore our Universe by creating synthetic models inside of our computer. We could be living in someone's computer simulation of our Universe.

Luke: Hmmm. And how did our programmer know how to set the laws of physics within their simulation to allow life forms to evolve and question their own existence?

Geraint: It's not that different from the multiverse hypothesis, actually. The great programmer may be exploring the effects of various parameters, running many individual simulations and examining the differing outcomes. The majority of simulations will result in dead universes, but we unsurprisingly find ourselves in one in which we can thrive.

Luke: So our glorious creator is a theoretical physicist who runs computer simulations of universes. That sounds a bit familiar . . . a god in our own image?

Anyway, it's a coherent idea. Let's walk around in its shoes for a while. Are there any interesting consequences of supposing that we are simulated beings on a simulated planet orbiting a simulated star in a simulated universe?

Geraint: Well, when we simulate physical phenomena, we need to make approximations. This is because we are limited by the computational power of the computers that we use. Consider a simulation of the evolution of matter in the Universe as it expands. While we know that gas is made of atoms, and we think that dark matter is constituted by some fundamental particle, there is no way that we can simulate the motions of individual atoms and particles. Today's computers just don't have the memory to cope with such immense numbers.

Cosmology is not alone in making such approximations. When we want to simulate water flowing through a pipe, or air over the wing of an aircraft, we do not treat the water and air as being composed of individual atoms, but instead as a continuous fluid. Why? Because these approximations work well in representing the real world on the scales we are interested in, even if they fail on the smallest scales.

Luke: So, if we are a simulation, are we an approximation of the more complex universe of the programmer?

Geraint: Not necessarily. The programmer could simulate any mathematically consistent physical law that they can dream up. Within our Universe, proposed physical laws are compared to nature, to see whether they provide an accurate description of observed phenomena. To do this, we need to examine their consequences. We can do this for any law, and we can even investigate laws for the sake of curiosity, or fun.

Unlike most pen and paper calculations, computer simulations can investigate universes by recreating them to some degree of detail. These are little universes inside a computer.

Luke: Let's be clear about one thing: you're claiming that we should consider these simulated universes to be just as real as ours. In our Universe, they are just long strings of zeros and ones. But someone 'living' inside the computer would experience a fully functional universe.

Geraint: I guess so. We don't know much about what physical properties are needed to make something conscious, but we assume that we can simulate them. We could – in principle! – simulate a brain, for example.

There is one more thing I think we need to consider. We've both written programs that use pseudo-random number generators to explore a set of possibilities. All kinds of simulated universes are produced for a variety of reasons. It might not be the case that the programmer's attention was on this universe alone, or in producing life. We could be a byproduct. The attention of the programmer might be on some other aspect of their simulation. They might not even know that we are here!

Luke: Well, I hope they don't get bored and stop the simulation.

Geraint: I guess we also have to hope that the cleaner in this higher universe does not unplug the computer to plug in their multi-dimensional vacuum cleaner.

Luke: I melted three logic boards last year – a new record – so I hope those higher-dimensional cooling fans are at maximum.

Those worries aside, suppose that the programmer runs their simulation with the intention of creating a universe that life forms can inhabit, and they know we are here and are monitoring our progress. Is this really that different from the God hypothesis?

Geraint: There are certainly similarities – a universe is established according to a plan and set running. But the intention in creating the Universe would be quite different. Instead of a Universe crafted for us by a god who is necessary and omniscient and morally perfect and all that, we could be little more than a high-school science project of a spotty teenager in a universe much more complex than our own.

Luke: At face value, we're just moving the problem up a level. Why does this universe permit life? Because it is the product of a universe that permits life, and computers and simulations. Guess what my next question is?

In fact, it seems worse than the multiverse. The multiverse involves generating a vast horde of universes, in which life will form somewhere, given enough variation. This simulated multiverse requires some kind of programmer and computer from the beginning.

Geraint: Point taken, but there is a way to make this work. Most of the possible universes that we have explored are – understandably – somewhat similar to ours. The equations of our Universe have received the most attention from physicists, and so are the ones that we're best able to solve.

What if, contrary to expectations, there is a distant oasis in possible physics space, with life-permitting universes as far as the eye can see. At least locally, life would not appear to be fine-tuned and perhaps, therefore, not too surprising. The simulation multiverse – if that's what you want to call it – shows how to start with *any* life-permitting universe and produce a multitude of other life-permitting universes, even fine-tuned ones.

Luke: Is there any way that we could tell if we were really living in a simulation?

Geraint: We could look for evidence that the Universe is discrete rather than continuous. This would look like small departures from our continuous laws of nature.

Also, anyone who has written a computer program will tell you that it is rarely perfect. Bugs and glitches sneak in, and unintended consequences sneak out. Maybe if we keep our eyes open for such glitches, we could work out whether this is a real universe or not. It won't be easy – without knowing how the code works, how do we recognize a glitch? What if a glitch is one of the imperfections of the Universe that we've already noted? What if, for example, the cosmological constant's extremely small value is just a rounding error, that is, the result of a computer not quite being able to carry all the digits?[32]

THE GOING DOWN OF THE SUN

Narrator: Our cosmologists are tired. It is late, and some weird ideas are starting to look far too plausible. It's time to come to some conclusions before the Sun sets on another Sydney day.

Luke: Well, that escalated quickly. I mean, that really got out of hand fast. We were just innocently wondering what would happen if the Universe were different. We ended up wandering through most of physics and cosmology, stopped by probability theory, dipped into mathematics, stumbled inexpertly through some philosophy, and even tried to put a multiverse in a computer.

I'll try to wrap up. A set of equations summarizes our best understanding of how nature works. We strongly suspect that these equations are not the final word in physics, but instead are a special case of some deeper, simpler, ultimate law. One aspect of our laws that looks

---

[32] Beane, Davoudi and Savage (2014) explore this idea, and a number of other consequences of the hypothesis that the Universe is a computer simulation. In fact, Konrad Zuse discussed the universe as a simulation in 1969. For more details, Fredkin (1990), Wolfram (2002, p. 1197) and 't Hooft (2013).

particularly incomplete is the presence of numbers – free parameters – in the equations themselves and in the solutions to these equations.

Looking for clues to new ideas, physicists have considered the consequences of varying the values of these parameters. The result is often swift disaster: seemingly small changes in the parameters lead to dramatic, uncompensated and detrimental changes in a universe's ability to create and support the complexity needed by life. This is called the *fine-tuning of the Universe for life*.

What might a deeper theory of physics tell us about these parameters? They could be mathematical constants, in which case they are hard-wired into the theory, and cannot be changed without changing the form of the equation or some other feature of the mathematical structure of the theory. They could also be related to a dynamical entity like a field, which changes with space and time. In this case, fine-tuning can furnish a completely different solution: multiverse + anthropic principle. There is a huge ensemble of universes, in which life is likely to form somewhere, and observers will only see universes that can create observers. This multiverse could even be the simulation of another life form in the next universe up.

Alternatively, all this talk of the ultimate laws of nature and the deepest level of explanation that science can hope to provide sends us towards some famously 'big' questions: why does anything at all exist? Why does this particular universe exist? According to naturalism, these questions are *unanswerable* – you'll just have to convince yourself that you didn't *really* want an answer anyway. According to theism, a *perfect* being can answer these deep questions. Its necessary existence explains why anything exists. Its moral perfection makes its creation of a moral-agent-permitting Universe like ours more likely. However, this does raise more questions! For example, does the idea of a necessary being even make sense?

Geraint: I think I still stick with the multiverse. Ours is but one of a vast sea of universes, and each with differing laws of physics and properties of matter, set at their birth through some cosmic roll of the dice. As we have seen, almost all of these universes are sterile, empty or soon-to-collapse. But, of course, we find ourselves in one of the extremely few universes that can support life – the anthropic principle in action.

Luke: The usual complaint about the multiverse is that it is extravagant and untestable. I suspect that this is a product of the difficultly of our cosmological theories. Extracting their predictions is hard. I think that it would be an extraordinary achievement for a multiverse theory to correctly and naturally predict the values of the constants and initial conditions of our Universe, and (in so doing) avoid the Boltzmann Brain problem.

But those Boltzmann Brains are a worry. A multiverse that avoids them must favour biological life forms – like us – over isolated, freak observers. This might sink the whole idea.

Geraint: I want to get some decent sleep tonight! Is there a rosier picture that I can fill my head with as I drop off?

Luke: I'm not sure about rosy, but here's another idea. The ability of our Universe to support life is partly a consequence of its ability to create stars. And when a massive star collapses at the end of its relatively short life, it can explode violently as a supernova. This explosion can be so violent that the core of the star is squeezed into unimaginable densities, forming a black hole. Lee Smolin (1997) has theorized – which in physics sometimes means 'guessed with equations' – that such a formation of a black hole could lead to the creation of a new baby universe that resembles its parent.

Geraint: Really? Creating a black hole actually creates a whole new universe?

Luke: That's the idea. Once you have a universe that can create stars and black holes, it starts to create baby universes that are just like it, with stars and black holes. And the formation of black holes in the baby universes creates even more universes, and then more and then more as we consider subsequent generations. We have reproducing universes!

If you think about this in the context of the multiverse, this would mean that through these subsequent generations, universes with stars and black holes would come to dominate the population of universes. Life may be able to find a place in these universes, being powered by the stars. In this way, life-bearing universes come to dominate the multiverse.

Geraint: And then life-bearing universes would be the norm, rather than a rare fluke.

Luke: Right. And while you still need to conjure up the first universe that can produce black holes, at least they seem easy to make – just let gravity do its thing.

In fact, the easiness of black holes might sink the idea. It's probably easier just to create black holes directly in a lumpy Big Bang or by fluctuations in an inflating universe rather than go to all the bother of creating stars. And, the physics underlying the idea is speculative, to say the least.

What about your simulated universe? We're inside an elaborate computer simulation, a synthetic universe created by some N-dimensional being messing about on a computer. Does that picture appeal to you at all?

Geraint: Not really. I find it neither comforting nor uncomfortable. Life is for living, no matter who or what brought it into being.

Narrator: At this, our cosmologists fall quiet and look around the park. They notice that the Sun is dipping well below the horizon, with a cold chill accompanying the encroaching darkness, and the park has now emptied of families. Our two cosmologists decide that it's time to conclude their discussions.

Geraint: How will we ever know what the answer is?

Luke: Well, there are a few barriers. A better understanding of the laws of nature, and new observational methods using neutrinos or gravitational waves, could let us see further back to the beginning.

Geraint: Right. We need that theory of quantum gravity. The other forces must have played an important role in the early evolution of the Universe, but the mathematics of General Relativity and quantum mechanics simply don't work with each other. And nobody knows the correct way of uniting them.

What about string theory, or quantum loop gravity, or one of the other cool things we see splashed across the pages of *New Scientist* and *Scientific American*?

Luke: Yes, there's been a lot of activity in these areas, but nothing by way of experimental evidence. The initial optimism about string theory is waning. The theory of everything has been ten years away for quite a while.

Geraint: What if there isn't a theory of everything?

Luke: Funnily enough, I think that we can make a crude anthropic argument here. The laws of nature seem to be more simple and elegant than life needs. We can easily make the laws of nature uglier in ways that would not affect the Universe's ability to create life. However, our Universe does not have these ugly features. This suggests that the simplicity we see in the laws of nature goes all the way down. There is reason to hope for a theory of everything.

One major worry is that we can't discover the right law because we don't yet speak the right mathematical language. But mathematics is pure thought – we can always think harder.

Geraint: If we don't have a theory of everything, can we peer 'through' the Big Bang and work out where the Universe came from?

Luke: We don't know, but it seems unlikely. We will see to as far as our approximations are valid, and no further.

In the meantime, any number of ways to test our ideas might remain stubbornly beyond our reach. We're left with incomplete evidence. Try to think clearly, consider all the options, and be mindful of your own biases.

Geraint: Well, that gives me something to think about.

Luke: It certainly does, but not tonight. It's getting late and I think it's about time for me to head home. Irrespective of how many other dead and sterile universes are out there, in this one I have a pair of kids that need a bath.

Geraint: Yeah. At the moment, this Universe is certainly enough to deal with! Goodnight.

Narrator: Our two cosmologists head off in different directions, slowly being swallowed up by the darkness.

Further Reading

There is an extensive literature on the fine-tuning of the Universe, from popular to professional.

BOOKS: POPULAR LEVEL

Just Six Numbers by Martin Rees. Highly recommended, with a strong focus on cosmology and astrophysics, as you'd expect from the Astronomer Royal. Rees gives a clear exposition of modern cosmology, including inflation, and concludes with a cogent defence of the multiverse.

The Goldilocks Enigma by Paul Davies. Davies is an excellent writer and has long been an important contributor to this field. His discussion of the physics is very good, especially the Higgs mechanism. When he strays into metaphysics, he is thorough and thoughtful.

The Cosmic Landscape: String Theory and the Illusion of Intelligent Design by Leonard Susskind. Susskind is a wonderful explainer, as his many online lectures will attest.

Constants of Nature by John Barrow. A discussion of the physics behind the constants of nature. An excellent presentation of modern physics, cosmology and their relationship to mathematics, which includes a chapter on the anthropic principle and a discussion of the multiverse.

Cosmology: The Science of the Universe by Edward Harrison. One of the best introductions to cosmology. The entire book is worth reading, not just the sections on life in the Universe and the multiverse.

At Home in the Universe by John Wheeler. A thoughtful and entertaining collection of essays, some of which touch on anthropic matters.

The Ambidextrous Universe by Martin Gardner. A fascinating look at the importance of symmetry and asymmetry in the Universe.

The Fallacy of Fine-Tuning: Why the Universe is Not Designed For Us and God and the Multiverse by Victor Stenger. This book's antiparticle. Stenger defends the view that physics, without needing a multiverse, has already solved all the problems of fine-tuning. Luke's review paper (below) is in part a response to this book, and finds many flaws.

BOOKS: ADVANCED

The Anthropic Cosmological Principle, by John Barrow and Frank Tipler. The standard in the field. Even if you can't follow the equations in the middle chapters, it's still worth a read as the discussion is quite clear. It becomes rather speculative in the final chapters, but it's obvious where to apply your grain of salt.

Universe or Multiverse, edited by Bernard Carr. A great collection of papers by most of the experts in the field.

SCIENTIFIC REVIEW ARTICLES

The field of fine-tuning grew out of the so-called 'large number hypothesis' of Paul Dirac, and related research by Hermann Weyl, Arthur Eddington, George Gamow and others. These discussions led to the recognition of fine-tuning when Robert Dicke explained the large number coincidences using the anthropic principle. Dicke's argument is examined and expanded in these *classic papers* of the field:

- Large Number Coincidences and the Anthropic Principle in Cosmology, Carter (1974).
- The Anthropic Principle and the Structure of the Physical World, Carr and Rees (1979).
- The Anthropic Principle, Davies (1983).

A number of papers, while not discussing fine-tuning, show how the macroscopic features of the Universe depend on the values of fundamental constants. They are great fun to work through if you don't mind a bit of maths.

- Dependence of Macrophysical Phenomena on the Values of the Fundamental Constants, Press and Lightman (1983).
- The Eighteen Arbitrary Parameters of the Standard Model in Your Everyday Life, Cahn (1998).

Here are a few good review papers, arranged in order of increasing technical level.

- Understanding the Fine Tuning in Our Universe, Cohen (2008). A nice introduction to the fine-tuning of nuclear binding and nucleosynthesis in stars, aimed at undergraduate physics students.
- Numerical Coincidences and 'Tuning' in Cosmology, Rees (2003).
- Why the Universe Is Just So, Hogan (2000). An excellent overview and update of the field, and one of the first papers to extend anthropic constraints to grand unified theories.
- The Fine-Tuning of the Universe for Intelligent Life, Barnes (2012). Luke's comprehensive review of the important work done in the field since Hogan's review in 2000.
- Life at the Interface of Particle Physics and String Theory, Schellekens (2013). A wide-ranging review from the standpoint of a string theorist.
- Varying Constants, Gravitation and Cosmology, Uzan (2011). The scientific field of fine-tuning overlaps with investigations of variations of fundamental constants in this Universe, for obvious reasons – both ask 'what would happen if the fundamental constants were different?' Uzan gives a thorough overview of this field.

PHILOSOPHICAL ARTICLES AND BOOKS

Universes by Leslie (1989) is a tremendously clear exposition of what conclusions we can and should draw from fine-tuning. More general reviews of some of the many philosophical issues raised by modern cosmology, including fine-tuning, are 'Issues in the Philosophy of Cosmology' by Ellis (2006) and 'Philosophy of Cosmology' by Smeenk (2013).

Part of the reason why the fine-tuning of the Universe for life is of such interest to philosophers is that it is often used as a premise in an argument for the existence of God. A lot of the literature on the fine-tuning argument, pro and con, misses the mark by a large margin, in our opinion. Some of the better ones include *Modern Physics and Ancient Faith* (Barr, 2003), *The Existence of God* (Swinburne, 2004), *Theism and Ultimate Explanation* (O'Connor, 2008), and 'The Teleological Argument: An Exploration of the Fine-Tuning of the Universe' (Collins, in Craig and Moreland, 2009).

Unsurprisingly, such claims have not gone unchallenged: *Logic and Theism* (Sobel, 2009), *Arguing about Gods* (Oppy, 2009), and 'Does the Universe Need God?' (Carroll, 2012).

References

Adams, Fred C. (2008). Stars in Other Universes: Stellar Structure with Different Fundamental Constants. *Journal of Cosmology and Astroparticle Physics*, **08**, 010.

Adams, Fred C. and Gregory Laughlin (1997). A Dying Universe: The Long Term Fate and Evolution of Astrophysical Objects. *Reviews of Modern Physics*, **69**(2), 337–372.

Aguirre, Anthony (2001). Cold Big-Bang Cosmology as a Counterexample to Several Anthropic Arguments. *Physical Review D*, **64**, 083508.

Albert, David Z. (2000). *Time and Chance*. Cambridge, MA; London: Harvard University Press.

Albrecht, Andreas and Lorenzo Sorbo (2004). Can the Universe Afford Inflation? *Physical Review D*, **70**, 063528.

Aldrich, John (2008). R. A. Fisher on Bayes and Bayes' Theorem. *Bayesian Analysis*, **3**(1), 161–170.

Augustine (415). *De Genesi Ad Litteram*. Translated and annotated by John Hammond Taylor (1982). New York: The Newman Press.

Augustine (420). *Enchiridion: On Faith, Hope, and Love*. Translated by Albert C. Outler. Philadelphia: Westminster Press, 1955.

Banks, Tom (2012). The Top 10^{500} Reasons Not To Believe in the String Landscape. arXiv:1208.5715.

Barnes, Luke (2012). The Fine-Tuning of the Universe for Intelligent Life. *Publications of the Astronomical Society of Australia*, **29**(4), 529–564.

Barnes, Luke (2014). Cosmology Q & A. *Australian Physics*, **51**, 42–46.

Barr, Stephen M. (2003). *Modern Physics and Ancient Faith*. Notre Dame, IN: University of Notre Dame Press.

Barr, Stephen M. and Almas Khan (2007). Anthropic Tuning of the Weak Scale and of m_u/m_d in Two-Higgs-Doublet Models. *Physical Review D*, **76**, 045002.

Barrow, John (2002). *The Constants of Nature: The Numbers That Encode the Deepest Secrets of the Universe*. London: Pantheon Books.

Barrow, John D. and Frank J. Tipler (1986). *The Anthropic Cosmological Principle*. Oxford: Clarendon Press.

Beane, Silas R., Zohreh Davoudi and Martin J. Savage (2014). Constraints on the Universe as a Numerical Simulation. *The European Physical Journal A*, **50** (148), 9–17.

Berry, Michael (1978). Regular and Irregular Motion. In S. Jorna, ed., *Topics in Nonlinear Dynamics*. New York: American Institute of Physics.

Boltzmann, Ludwig (1895). On Certain Questions of the Theory of Gases. *Nature*, **51**, 413–415.

Brandenberger, Robert H. (2008). String Gas Cosmology. arXiv:0808.0746.

Burgess, Cliff and Guy Moore (2007). *The Standard Model: A Primer*. Cambridge: Cambridge University Press.

Cahn, Robert (1996). The Eighteen Arbitrary Parameters of the Standard Model in Your Everyday Life. *Reviews of Modern Physics*, **68**, 951–959.

Carr, Bernard (ed.) (2009). *Universe or Multiverse?* Cambridge: Cambridge University Press.

Carr, Bernard J. and Martin J. Rees (1979). The Anthropic Principle and the Structure of the Physical World. *Nature*, **278**(12), 605–612.

Carroll, Lewis (1871). *Through the Looking-Glass, and What Alice Found There*. New York: Macmillan.

Carroll, Sean M. (2006). Is Our Universe Natural? *Nature*, **440**, 1132–1136.

Carroll, Sean M. (2010). *From Eternity to Here: The Quest for the Ultimate Theory of Time*. Oxford: Oneworld Publications.

Carroll, Sean M. (2012). Does the Universe Need God? In J. B. Stump and Alan G. Padgett, eds., *The Blackwell Companion to Science and Christianity*. Chichester: Wiley-Black.

Carroll, Sean M. (2014). In What Sense Is the Early Universe Fine-Tuned? arXiv:1406.3057.

Carroll, Sean M. and Heywood Tam (2010). Unitary Evolution and Cosmological Fine-Tuning. arXiv:1007.1417.

Carter, Brandon (1974). Large Number Coincidences and the Anthropic Principle in Cosmology. in M. S. Longair, ed., *Confrontation of Cosmological Theories With Observational Data*. Dordrecht: D. Reidel.

Cathcart, Brian (2004). *The Fly in the Cathedral: How a Small Group of Cambridge Scientists Won the Race to Split the Atom*. London: Viking.

Chalmers, David (1996). *The Conscious Mind: In Search of a Fundamental Theory*. New York: Oxford University Press.

Close, Frank (2011). *The Infinity Puzzle: Quantum Field Theory and the Hunt for an Orderly Universe*. New York: Basic Books.

Cohen, Bernard (2008). Understanding the Fine Tuning in Our Universe. *The Physics Teacher*. **46**, 285–289.

Cook, Matthew (2004). Universality in Elementary Cellular Automata. *Complex Systems*, **15**(1), 1–40.

Craig, William Lane and J. P. Moreland (eds.) (2009). *The Blackwell Companion to Natural Theology*. Oxford: Wiley-Blackwell.

Dass, Tulsi (2005). Measurements and Decoherence. arXiv:quant-ph/0505070.

Davies, Paul. C. W. (1983). The Anthropic Principle. *Progress in Particle and Nuclear Physics*, **10**, 1–38.

Davies, Paul. C. W. (2006). *The Goldilocks Enigma: Why Is the Universe Just Right for Life?* London: Allen Lane.

Davies, Paul. C. W. (2010). *The Eerie Silence*. Boston; New York: Houghton Mifflin Harcourt.

Dawkins, Richard (1995). Reply to Michael Poole. *Science and Christian Belief*, **7**(1), 48–49.

Dawkins, Richard (2006). *The God Delusion*. London: Bantam Press.

Diamond, Jared (2005). *Collapse: How Societies Choose to Fail or Survive*. London: Viking Penguin.

Dine, Michael and Alexander Kusenko (2003). The Origin of the Matter-Antimatter Asymmetry. *Reviews of Modern Physics*, **76**(1), 1–30.

Dingle, Herbert (1953). On Science and Modern Cosmology (Presidential Address). *Monthly Notices of the Royal Astronomical Society*, **113**, 393–407.

Eagle, Antony (2011). *Philosophy of Probability: Contemporary Readings*. London: Routledge.

Eddington, Arthur S. (1928). *The Nature of the Physical World*. Cambridge: Cambridge University Press.

Ellis, George F. R. (2007). Issues in the Philosophy of Cosmology. In Jeremy Butterfield and John Earman, eds., *Handbook in Philosophy of Physics*. Amsterdam: North Holland.

Epelbaum, Evgeny, Hermann Krebs, Dean Lee and Ulf-G. Meißner (2011). Ab Initio Calculation of the Hoyle State. *Physical Review Letters*, **106**(19), 192501.

Epelbaum, Evgeny, Hermann Krebs, Timo A. Lähde, Dean Lee and Ulf-G. Meißner (2013). Viability of Carbon-Based Life as a Function of the Light Quark Mass. *Physical Review Letters*, **110**(11), 112502.

Evrard, Guillaume and Peter Coles (1995). Getting the Measure of the Flatness Problem. *Classical and Quantum Gravity*, **12**(10), L93.

Feser, Edward (2009). *Aquinas (A Beginner's Guide)*. Oxford: Oneworld Publications.

Feynman, Richard P. (1965). *The Character of Physical Law*. Cambridge, MA: MIT Press.

Feynman, Richard P. (1988). *QED. The Strange Theory of Light and Matter.* Princeton, NJ: Princeton University Press.

Feynman, Richard, Robert B. Leighton and Matthew L. Sands (1970). *The Feynman Lectures on Physics (3 Volume Set).* Boston: Addison Wesley Longman.

Fowler, William A. (1966). The Stability of Supermassive Stars. *The Astrophysical Journal,* **144**, 180–200.

Fredkin, Edward (1990). Digital Mechanics: An Informational Process Based on Reversible Universal Cellular Automata. *Physica,* **D45**, 254–270.

Gamow, George (1965). *Mr. Tompkins in Paperback.* Cambridge: Cambridge University Press.

Gardner, Martin (1964). *The Ambidextrous Universe.* New York: Basic Books.

Gardner, Martin (1970). Mathematical Games: The Fantastic Combinations of John Conway's New Solitaire Game 'Life'. *Scientific American,* **223**, 120–123.

Gibbons, Gary W. and Neil Turok (2008). Measure Problem in Cosmology. *Physical Review D,* **77**, 6, 063516.

Gleick, James (2012). *The Information: A History, A Theory, A Flood.* New York: Vintage Books.

Goldacre, Ben (2009). *Bad Science: Quacks, Hacks, and Big Pharma Flacks.* London: Fourth Estate.

Goldacre, Ben (2014). *Bad Pharma: How Drug Companies Mislead Doctors and Harm Patients.* London: Faber & Faber.

Greene, Brian (1999). *The Elegant Universe.* New York: W. W. Norton.

Gribbin, John (1985). *In Search of the Double Helix.* New York: McGraw-Hill.

Gross, David J. (1996). The Role of Symmetry in Fundamental Physics. *Proceedings of the National Academy of Sciences USA,* **93**(25), 14256–14259.

Guy, R. K. (2008). John H. Conway. In D. J. Albers and G. L. Alexanderson, eds., *Mathematical People: Profiles And Interviews,* 2nd edn. Wellesley, Massachusetts: A K Peters, Ltd.

Hall, Lawrence J., David Pinner and Joshua T. Ruderman (2014). The Weak Scale from BBN. *Journal of High Energy Physics,* **12**(134), 29.

Harnik, Roni, Graham D. Kribs and Gilad Perez (2006). A Universe Without Weak Interactions. *Physical Review D,* **74**, 035006.

Harrison, Edward (2000). *Cosmology: The Science of the Universe.* Cambridge: Cambridge University Press.

Hawking, Stephen W. (1988). *A Brief History of Time: From the Big Bang to Black Holes.* New York: Bantam Books.

Hawking, Stephen W. and Don N. Page (1988). How Probable Is Inflation? *Nuclear Physics B,* **298**, 789–809.

Helbig, Phillip (2012). Is There a Flatness Problem in Classical Cosmology? *Monthly Notices of the Royal Astronomical Society*, **421**, 561–569.

Hogan, Craig J. (2000). Why the Universe Is Just So. *Reviews of Modern Physics*, **72**, 1149–1161.

Hogan, Craig J. (2009). Quarks, Electrons, and Atoms in Closely Related Universes. In Bernard Carr, ed., *Universe or Multiverse?* Cambridge: Cambridge University Press.

Hollands, Stefan and Robert M. Wald (2002). Essay: An Alternative to Inflation. *General Relativity and Gravitation*, **34**, 2043–2055.

Holt, Jim (2012). *Why Does The World Exist? One Man's Quests for the Big Answer*. London: Profile Books.

Hoyle, Fred (1950). *The Nature of the Universe*. Oxford: Blackwell.

Hoyle, Fred (1957). *The Black Cloud*. London: William Heinemann Ltd.

Hoyle, Fred (1994). *Home Is Where the Wind Blows: Chapters from a Cosmologist's Life*. California: University Science Books.

Hume, David (1779). *Dialogues Concerning Natural Religion*. Web edition published by eBooks@Adelaide: ebooks.adelaide.edu.au/h/hume/david/h92d

Jaynes, Edwin (2003). *Probability Theory: The Logic of Science*. Cambridge: Cambridge University Press.

Jeans, James (1931). *The Stars in Their Courses*. Cambridge: Cambridge University Press.

Kofman, Lev, Andrei Linde and Viatcheslav Mukhanov (2002). Inflationary Theory and Alternative Cosmology. *Journal of High Energy Physics*, **10**, 057.

Kragh, Helge (2010). When Is a Prediction Anthropic? Fred Hoyle and the 7.65 Mev Carbon Resonance. philsci-archive.pitt.edu/5332/

Kreeft, Peter (1990). *A Summa of the Summa*. San Francisco: Ignatius Press.

Leslie, John (1989). *Universes*. London: Routledge.

Li, Ming and Paul M. B. Vitányi (2008). *An Introduction to Kolmogorov Complexity and Its Applications*. New York: Springer-Verlag.

Livio, M., D. Hollowell, J. W. Truran and A. Weiss (1989). The Anthropic Significance of the Existence of an Excited State of C-12. *Nature*, **340**(6231), 281–284.

Loeb, Abraham (2014). The Habitable Epoch of the Early Universe. *International Journal of Astrobiology*, **13**(4), 337–339.

MacDonald, J. and D. J. Mullan (2009). Big Bang Nucleosynthesis: The Strong Nuclear Force Meets the Weak Anthropic Principle. *Physical Review D*, **80**(4), 043507.

McGrayne, Sharon (2012). *The Theory That Would Not Die*. New Haven, CT: Yale University Press.

McGrew, Timothy, Lydia McGrew and Eric Vestrup (2003). Probabilities and the Fine-Tuning Argument. In Neil Manson, ed., *God and Design*. London: Routledge.

Mears, Ray (2003). *The Real Heroes of Telemark*. London: Hodder & Stoughton.

Meißner, Ulf-G. (2015). Anthropic Considerations in Nuclear Physics. *Science Bulletin*, **60**(1), 43–54.

Mitton, Simon (2011). *Fred Hoyle: A Life in Science*. Cambridge: Cambridge University Press.

Nussbaumer, Harry and Lydia Bieri (2009). *Discovering the Expanding Universe*. Cambridge: Cambridge University Press.

O'Connor, Timothy (2008). *Theism and Ultimate Explanation*. London: Wiley-Blackwell.

Olive, K. A. et al. (Particle Data Group) (2014). Review of Particle Physics. *Chinese Physics C*, **38**, 090001.

Oppy, Graham (2009). *Arguing About Gods*. Cambridge: Cambridge University Press.

Osler, Margaret J. (2010). Myth 10. That the Scientific Revolution Liberated Science from Religion. In Ronald L. Numbers, ed., *Galileo Goes to Jail and Other Myths about Science and Religion*. Cambridge, MA: Harvard University Press.

Page, Don N. (1983). Inflation Does Not Explain Time Asymmetry. *Nature*, **304**, 39–41.

Penrose, Roger (1979). Singularities and Time Asymmetry. In W. Israel and S.W. Hawking, eds., *General Relativity: An Einstein Centenary Survey*. Cambridge: Cambridge University Press.

Penrose, Roger (2004). *The Road to Reality: A Complete Guide to the Laws of the Universe*. London: Vintage.

Pochet, T., J. M. Pearson, G. Beaudet and H. Reeves (1991). The Binding of Light Nuclei, and the Anthropic Principle. *Astronomy and Astrophysics*, **243**(1), 1–4.

Polchinski, Joseph (2006). The Cosmological Constant and the String Landscape. arXiv:hep-th/0603249.

Pratchett, Terry. (1994). *Soul Music*. London: Victor Gollancz.

Press, William H. and Alan P. Lightman (1983). Dependence of Macrophysical Phenomena on the Values of the Fundamental Constants. *Philosophical Transactions of the Royal Society A*, **310**(1512), 323–336.

Rees, Martin (2001). *Just Six Numbers: The Deep Forces That Shape The Universe*. New York: Basic Books.

Rees, Martin (2003). Numerical Coincidences and 'Tuning' in Cosmology. In N. C. Wickramasinghe, Geoffrey Burbidge and J. V. Narlikar, eds., *Fred Hoyle's Universe*. Dordrecht, The Netherlands: Kluwer.

Reia, Sandro and Osame Kinouchi (2014). Conway's Game of Life Is a Near-Critical Metastable State in the Multiverse of Cellular Automata. *Physical Review E*, **89** (5), 052123.

Rovelli, Carlo (2015). Aristotle's Physics: A Physicist's Look. *Journal of the American Philosophical Association*, **1**(01), 23–40.

Schellekens, A. N. (2013). Life at the Interface of Particle Physics and String Theory. *Reviews of Modern Physics*, **85**, 1491–1540.

Schilpp, P. (ed.) (1969). *Albert Einstein: Philosopher-Scientist*. Peru, IL: Open Court Press.

Schneider, Nathan (2013). *God in Proof: The Story of a Search from the Ancients to the Internet*. Berkeley: University of California Press.

Schrödinger, Erwin (1935). Die gegenwärtige Situation in der Quantenmechanik (The Present Situation in Quantum Mechanics). *Naturwissenschaften*, **23**(49), 823–828.

Seuss, Dr (1954). *Horton Hears a Who!* New York: Random House.

Silver, Nate (2015). *The Signal and the Noise*. New York: Penguin Books.

Smeenk, C. (2013). Philosophy of Cosmology. In R. Batterman, ed., *Oxford Handbook of Philosophy of Physics*. New York: Oxford University Press.

Smolin, Lee (1997). *The Life of The Cosmos*. New York: Oxford University Press.

Sobel, Jordan Howard (2009). *Logic and Theism*. Cambridge: Cambridge University Press.

Stenger, Victor (2011). *The Fallacy of Fine-Tuning: Why the Universe Is Not Designed for Us*. New York: Prometheus Books.

Storrie-Lombardi, Lisa J. and Arthur M. Wolfe (2000). Surveys for $z > 3$ Damped Lyα Absorption Systems: The Evolution of Neutral Gas. *The Astrophysical Journal*, **543**(2), 552–576.

Susskind, Leonard (2005). *The Cosmic Landscape: String Theory and the Illusion of Intelligent Design*. New York: Little, Brown and Company.

Susskind, Leonard and George Hrabovsky (2014). *Classical Mechanics: The Theoretical Minimum*. London, Penguin Books.

Swinburne, Richard (2004). *The Existence of God*. Oxford: Oxford University Press.

't Hooft, Gerard (2013). Duality Between a Deterministic Cellular Automaton and a Bosonic Quantum Field Theory in 1+1 Dimensions. *Foundations of Physics*, **43**(5), 597–614.

Taleb, Nassim Nicholas (2010). *The Black Swan*. New York: Random House.

Tegmark, Max (1997). Letter to the Editor: On the Dimensionality of Spacetime. *Classical and Quantum Gravity*, **14**(4), L69–75.

Tegmark, Max. (1998). Is 'The Theory of Everything' Merely the Ultimate Ensemble Theory? *Annals of Physics*, **270**(1), 1–51.

Tegmark, Max and Martin J. Rees (1998). Why Is the Cosmic Microwave Background Fluctuation Level 10^{-5}? *The Astrophysical Journal*, **499**(2), 526–532.

Tegmark, Max, Alexander Vilenkin and Levon Pogosian (2005). Anthropic Predictions for Neutrino Masses. *Physical Review D*, **71**(10), 103523.

Tegmark, Max, Anthony Aguirre, Martin Rees and Frank Wilczek (2006). Dimensionless Constants, Cosmology, and Other Dark Matters. *Physical Review D*, **73**(2), 023505.

Turok, Neil (2002). A Critical Review of Inflation. *Classical and Quantum Gravity*, **19**, 3449.

Uzan, Jean-Philippe (2011). Varying Constants, Gravitation and Cosmology. *Living Reviews in Relativity*, **14**(2).

Vallentin, Antonina (1954). *Einstein: A Biography*. Translated from the French by Moura Budberg. London: Weidenfeld and Nicolson.

Ward, Keith (2009). *Why There Almost Certainly Is a God*. Oxford: Lion Hudson.

Way, M. J. and D. Hunter (eds.) (2013). *Origins of the Expanding Universe: 1912–1932*. ASP Conference Series 471, San Francisco: Astronomical Society of the Pacific.

Weinberg, Steven (1987). Anthropic Bound on the Cosmological Constant. *Physical Review Letters*, **59**, 2607–2610.

Weinberg, Steven (1993). *Dreams of a Final Theory*. London: Vintage Books.

Weinert, Friedel (2004). *The Scientist as Philosopher: Philosophical Consequences of Great Scientific Discoveries*. New York: Springer.

Wheeler, John (1994). *At Home in the Universe*. New York: American Institute of Physics.

White, John D. (1979). God and Necessity. *International Journal for Philosophy of Religion*, **10**, 177.

Williams, Bernard (1978). *Descartes: The Project of Pure Reason*. New York: Penguin.

Winsberg, Eric (2012). Bumps on the Road to Here (from Eternity). *Entropy*, **14**(3), 390–406.

Wolfram, Stephen (1984). Universality and Complexity in Cellular Automata. *Physica D: Nonlinear Phenomena*, **10**(2), 1–35.

Wolfram, Stephen (2002). *A New Kind of Science*. Champaign, IL: Wolfram Media.

Zurek, Wojciech H. (2002). Decoherence and the Transition from Quantum to Classical: Revisited. arXiv:quant-ph/0306072.

Zuse, Konrad (1969). *Rechnender Raum (Calculating Space)*. Braunschweig: Friedrich Vieweg & Sohn.

Index

Printed in the United States
By Bookmasters